Ivory Tower, House of Cards:
How Scholars and their Publishers Violate Science

John Major Jenkins

This book is a narrative of over two decades of exchanges with scholars, their academic publishers and employers, and the professional association that oversees and validates them. It focuses on recent exchanges, largely in 2015, and documents a series of officially filed error lists and complaints regarding scholarly errors needing correction, and the responses of the publishers, the science agency (NASA), and committees (the AAUP) that are appointed to oversee and uphold academic standards. A bizarre world of contradictions, evasions, bigotry, and sanctioned character assassination is exposed, indicting an elite club of Ivory Tower scholars, friends and colleagues engaged in sloppy research and guild protection whose behavior violates science and threatens their bastion of unethical self-interest with immanent collapse, like a flimsy house of cards.

This book is essentially a journalistic investigation and exposé of taxpayer-supported government scientists, professional scholars and educators, and university press publishers. Many of these people and institutions have been supported by grants and funds and therefore must be willing to have their words and deeds critiqued. Proceeds from this book will be donated to *The Maya Conservancy*, a non-profit that works with the contemporary Maya in the highlands of Guatemala.

Ivory Tower, House of Cards: How Scholars and their Publishers Violate Science. John Major Jenkins © 2017.

ISBN 978-0-9985868-2-3

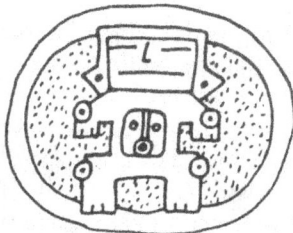

Four Ahau Press

"I've surveyed and critiqued 2012 authors and their various theories quite thoroughly, and objectively observe that none of it is very compelling, in terms of figuring out what the ancient creators of the 2012 calendar (the Long Count) thought about 2012. My own efforts took the rather obvious approach of asking where and when the Long Count was invented, and who did it. I was thus led to the pre-Classic site of Izapa, to the ballgame, and the Maya Creation Myth. What I found, in the evidence, was startling and it meant the ancient skywatchers of Mesoamerica were much more advanced than previously expected. For they could calculate solar alignments within the great cycle of the precession of the equinoxes, and formulated a profound cosmology around what that means. And the anchor of their system points to our times, to December 21, 2012."

—John Major Jenkins, 2009, Notebook #68

Table of Contents

Online Appendices, Contents:

Appendix 1: The Rest of the Iceberg
    a. Stuart-Houston
    b. Campion
    c. Van Stone
    d. Stan Guenter
    e. Review of MacLeod & Van Stone's "Great Return" article
    f. Review of Michael Grofe's 2003 article
    g. Various Others (11 topics)

Appendix 2: Additional Online Resources
    a. Annotated List of Links
    b. The Center for 2012 Studies
    c. Update2012.com

Appendix 3: Verbatim Filed Complaints and Emails

First Preface

It is November 1, the Samhain cross-quarter of the year 2015. The wheel has turned and the veil has thinned. Ghosts and spirits hover, felt but unseen. It is time to lay them to rest, with truth and light. And so I offer this book, to shed light on the spooks who inhabit our world, who deceive and lie and distract us from the truth.

Early November rings with memories in my personal past. Times of endings and new beginnings. In particular, it was exactly nineteen years ago — one Metonic Cycle — that I completed many years of research into Maya cosmology and 2012. The centerpiece of my book *Maya Cosmogenesis 2012* was the section on Izapa, which was a self-standing monograph. With a sense of accomplishment, I dated the completed study: November 7, 1996. The following day was a Friday — the date of a party with some friends. And there I began my relationship with the woman who would become my wife.

Our many years together were filled with sharing, love, laughter, tears and deep recognition. The marriage lasted fourteen years, and we were together for sixteen. Two cycles of Venus-Sun. We built our two homes with creativity and consciousness, tried to have kids, accepted the reality of it not happening, loved and cared for our two cats, traveled together and experienced many wonderful things. The last time I saw her was on her 52nd birthday, by her side at her deathbed, nine days before she transitioned. Her last words to me were "I'll see you later, stay in touch."

During her treatments, which stretched over almost a year, we had many conversations, renewing that special deep connection we had, about spiritual things. A part of her could easily open up to a kind of angelic certainty about the ways of the world, a knowing without judgment. She once said that these times, in this world, were filled with negativity and corruption, and she wasn't afraid of death. She knew she was going to a better place, where a purity of love and truth existed. "Yes," I said, "I know what you mean." And I meant it.

My lovely wife had to endure much struggle in her own life, and also had to live and feel my own struggles in my writing and teaching career. Much pain and difficulty resulted from attacks on me personally and my professional work by corrupt and negative scholars, compounded by the media's idiocy about the 2012 topic. The deceit, betrayals of trust, and outright maliciousness were often hard to handle. It was, however, easy to see that many of my detractors were motivated, in their bad behavior, by typical human flaws and vices — jealousy primarily. My Ellie believed in me and my work, and her love and presence sustained me through years of triumphs, struggles, breakdowns and breakthroughs. I will always be grateful for her special soul-presence in my life, and hold dear the love we shared.

I know she would wish for me to move on, but I also know that she wanted the corrupt critics and malicious scholars to be exposed. Thus, this book is a necessary capstone to conclude this work. It signals the end of a phase of my work and opens the door on a new beginning, just as occurred 19 years ago. This book is necessary to clear the slate and tell the truth, to expose those who have poisoned the 2012 discussion with lies, deception, and — most disappointingly — a failure to do their jobs as reputable scientists. It's all here; there's nothing more that need be said. I suspect that contemporary readers, journalists, and social commentators won't care much about this story. It's probably something that only future historians will understand in its proper context and perspective.

I am now approaching my own Calendar Round milestone — my 52nd solar circuit will soon be completed. This, too, offers an end-beginning nexus. There is much more to explore and experience in this life. My new interests have been developing and building steam for a few years now. They aren't exactly a radical departure from my thirty years of engagement with Maya culture, shamanism, calendrics, and cosmology. Rather, they tap into a deeper strata from which that work sprang. Deeper currents have been identified which embrace my own ancestral background and deeply felt poetic calling, sending me back through family origins among Celtic sages and magicians in Wales and Ireland. It takes me beyond mere genealogy, into a larger field of cultural identity and belonging. Embracing ones deep, indigenous, ancestral soul is to return to the center, to the wellspring of life that washes and renews.

At the temple altar beside that inner fountain, I offer and release this work, with my ever-present intention of serving truth, honesty, and clarity.

John Major Jenkins
November 1, 2015

❖ ❖ ❖ ❖ ❖

Second Preface

Since the completion of this book in late 2015, Maya scholar Anthony Aveni released another book that critiqued my work and 2012 ideas. Titled *Apocalyptic Anxiety*, it was also published by the University Press of Colorado (as was his previous book) and was released in May of 2016. I immediately identified factual errors as well as the usual blend of snide loaded lingo, guilt by association tactics, and ad hominem jabs. I once again engaged Aveni and his publisher in order to seek errata corrections.

The process quickly looked like it was going down the same road as my previous effort (documented in the current book), so by the end of June I had composed a detailed exposé of the unremitting corruption in academic publishing. This 29,000-word document is called "The End of an Error: The Cure for Aveni's Apocalyptic Anxiety" and is posted for free online at www.Update2012.com/Review-of-Aveni2016.pdf. But my efforts continued to unfold and after a phone call with Darrin Pratt, the University Press of Colorado's Director, the attitude shifted and I found that Aveni began to acknowledge, rather than deny, the errors. I suspect he was instructed to behave like a responsible scholar. This truth telling process was not without delays and resistance, and in fact it took over three months to reach closure. But the end result was that all of the errors I pointed out were acknowledged and corrected for a second printing of the book. All remaining copies of the error-containing first printing were pulped. The eBook was corrected, and in the front matter it was stated that this was a "corrected second printing."

So, a victory had finally occurred. I consequently composed another review, shorter in length than the previous one, which documented the corrections. This new review, titled "Correcting Aveni's 2016 book *Apocalyptic Anxiety*," has been added to the present book as an addendum.

I decided to release the main section of the book on Amazon Create Space for the benefit of those who wish to understand the trials and tribulations that those who speak truth to a corrupt Ivory Tower have to go through, in order to get basic factual errors corrected in peer reviewed university press books. My purpose is to benefit progress in Maya Studies and not personal profit. Therefore, profits from this book will be donated to *The Maya Conservancy*, a non-profit that works with the contemporary Maya in Highland Guatemala. Similarly, the extensive appendices which provide all the detailed documentation which underlies this work, have been posted for free online at www.Update2012.com, also linked at *The Center for 2012 Studies* (www.thecenterfor2012studies.com).

Recent months have found me in a health crisis, having been diagnosed with Stage IV kidney cancer, so I have stepped up my efforts to publish my various manuscripts and unreleased projects. These include my novel, *Three Plumes of Judas*, and a re-release of my 1989 travelog, *Journey to the Mayan Underworld*. Both can be found on Amazon and Create Space.

John Major Jenkins
February 13, 2017

Chapter 1. The Framework of Denial, 1966-2015

In 1966 an iconic book called *The Maya* was published. Written by Michael Coe, an established Maya archaeologist, it is still in print today and has gone through nine editions. It was distinguished by the fact that it contained a few sentences that mentioned, for the first time in the published record, the infamous 2012 date of the Maya calendar.[1] The book was published with Thames & Hudson, a respected trade publisher in England, and was oriented to a mainstream audience. By today's standards it's quite academic in tone and content. In his chapter on the Mesoamerican calendar, written as a general introduction for his readers, Coe wrote:

> There is a suggestion that each of these [time periods] measured thirteen Baktuns, or something less than 5200 years, and that Armageddon would overtake the degenerate peoples of the world and all creation on the final day of the thirteenth. ... Our present universe ... is to be annihilated ... when the Great Cycle of the Long Count reaches completion (Coe 1966: 174).

Here we have, packaged with the first mention of the date, a curious assertion that the date signifies a Maya "Armageddon." Later, Coe and his defenders have explained that this was just a tongue-in-cheek sort of throwaway line. Perhaps a bit of a joke to shock his readers.[2]

Coe was never particularly interested in the unfolding 2012 literature. He claims he was unaware of Frank Waters' popular 1975 book *Mexico Mystique* (which adopted Coe's incorrect calculation of the end-date, as December 24, 2011) and his interests have led him to other areas of archaeological investigation (for example, Cambodia). But he was tapped by fate or destiny to note the date and be the first to articulate a guess at what the Maya thought about 2012. His words are where the doomsday-2012 meme began, although years later Maya scholar David Stuart blamed "New Age hacks."[3]

Another scientist, anti-2012 crusader David Morrison at NASA, asserted that I was the origin of the Maya calendar doomsday meme, writing "The claims about the Mayan calendar predicting catastrophe in 2012 began with John Major Jenkins (a self-styled 'independent researcher') in the early 1990s" (see Chapter 3a for

---

[1] Coe used the correct correlation family, the so-called GMT correlation, but in applying it he miscalculated the end date to December 24, 2011. This was picked up by Frank Waters and others, but was eventually corrected in later editions of his book. However in the latest 9th edition, he switched to the "286" correlation. Only the "283" correlation makes the cycle-ending fall on December 21, 2012, and all the other proposals do not pass the test of the interdisciplinary criteria. See www.thecenterfor2012studies.com/2012center-note18.pdf.

[2] In his book Stuart gives Coe a pass on this issue, saying it was a speculative aside (2011:305). See also Coe (2011) in Joseph Gelfer (2011). See also Note 4.

[3] See Stuart's "2012 Q & A" on his Maya Decipherment blog (2009). I attempted a correction to Stuart's false statement on the Aztlan e-list, but I was admonished by the moderator and my post was censored.

details). Other scholars (e.g., Ed Krupp) crafted their own clever versions of this utterly false and denigrating position.

Through eight editions, all offering revisions, Dr. Coe never altered or commented further on his Armageddon statement. In 2011 he wrote, on the eve of the publication of the 8th edition, that he still saw no reason to change it.[4] But then, lo and behold, the 9th edition was released in May of 2015 and big changes were afoot. Coe, now in his 80s, enlisted the help of another scholar, Stephen Houston, and the book was redesigned with many revisions. The Armageddon statement was removed, without comment, and was replaced with new words on 2012:

> The problem is that the events of 4 Ajaw 8 Kumk'u, although esoteric and difficult to interpret, seem benign rather than catastrophic. ... These are momentous yet hardly horrific events. The few Classic-period citations of 24 December AD 2012, tend, if anything, to be rather dull. They note the expected completion of a cycle but nothing like the prophecies found in Colonial Maya books or among the Aztecs (Coe & Houston 2015: 249-250, 9th edition of Coe 1966).

So, Coe's book, from 1966 to 2015, in its 1st and 9th editions, have brought us from 2012 being an "Armageddon" — the most dramatic and destructive scenario imaginable — to 2012 being, if anything, "rather dull." Otherwise, it's not "anything" at all, simply nothing. The importance of *nothing* is illustrated in what is *not reported* in these academic dismissals of 2012, in that certain ideas and breakthroughs were ignored and treated as of they were nothing. We hear nothing, for example, in what could have been a one-page survey of the literature and ideas published *by Coe's and Houston's colleagues*, regarding the many articles and books published on the 2012 topic between 2008 and 2014. We hear nothing of my prominent and pioneering work, or the controversies it triggered among the status quo, high up in their Ivory Tower. We do get a brief allusion to the two 2012 inscriptions,[5] but they were passed over quickly as if they were largely irrelevant.

For years, in the 1990s up to 2006, scholars said 2012 was irrelevant because we had no inscriptions mentioning the date. Then, Tortuguero Monument 6 became widely known. What of it? Not really that important, because it's only one date. Then the La Corona 2012 inscription was discovered in early 2012. What of it? According to Coe and Houston, nothing much to speak of. This is how it happens in academia. The process of "isolated sequential dismissal" allows the weight of accumulated evidence to be ignored. Those two inscriptions provided evidence for how the ancient Maya thought about 2012. But for years scholars (particularly Stephen Houston and David Stuart) have asserted that 2012 didn't mean much of anything to the Maya. That bias was adamantly maintained up through 2012, and now to 2015, despite the compelling evidence of these two inscriptions.

The first and last editions of Coe's classic tome bookends the Alpha and the Omega of scholarly attitudes toward 2012. We go from one extreme to the other.

---

[4] In Coe's preface to Gelfer (2011, *Deconstructing the Countercultural Apocalypse*).

[5] Tortuguero Monument 6, which came to widespread attention in 2006 (and which Houston previously wrote about), and La Corona Block 5, discovered in early 2012.

We find two opposite but equally undiscerning and unsupportable positions on 2012. Inside of these bookends, on a shelf 49 years long, everything of interest in 2012 research and discovery occurred, but none of it was cited, quoted, or discussed in any of the nine editions. This should be, at the very least, perplexing for any person observing the situation, professional scholar or not. Coe and Houston illustrate in stark terms why 2012, as a topic of rational investigation, has suffered in the hands of professional scholars. They wanly dismiss it or joke about it, like Harvard frat boys around a cooler (or keg), but they have been extremely resistant to rationally investigating it. Moreover, many prominent Mayanists would do much more than this — they would become hostile toward 2012 and its nefarious "proponents." It is within this irrational echo chamber and under the shadow of such academic bias and malpractice that my lifework has unfolded, and which my Experiment now addresses.

The Experiment alluded to in the title of Chapter 3 provides the guts of this book. I am asking the question: is academic publishing broken? In other words, are the checks-and-balances and the principles of error recognition and correction that are held so high, as a fundamental value of reputable scholarship, in actual fact being practiced? My Experiment asks professional scholars, the scientific institutions that employ them, and their academic publishers to apply *their own stated policies* in assessing a number of selected statements and publications. The conclusions of this experiment will be drawn from the resulting data.

There would be no point in conducting this experiment on a topic that was already established within the consensus of academia. Indeed, in that arena there would be little in the way of dissent, defensive sloppiness, shooting-from-the-hip critiques, *ad hominem* character assassinations, or controversy to produce anything that needed correcting. The orthodoxy would be established and unquestioned, with no outside pressure on the orthodoxy caused by new discoveries. No, my experiment must test academia precisely where it is suppose to do its job — the rational investigation and testing of new theories and ideas, and the resulting reconfiguration of, or abandonment of, old models and attitudes that no longer are supportable by the evidence. That's how science works. At least, that's how science *should* work, if it doesn't get violated by those who are trusted and ordained to practice it.

So, the perfect battleground for the Experiment is my area of expertise, a field of study I've been aware of and exploring in one form or another for almost forty years.[6] That is: 2012. Right away, many readers will probably roll their eyes and project all kinds of judgment and bias into the very mention of this concept. 2012? I mean doomsday, right? No, I don't mean doomsday. I mean, and have always meant, the 2012 date as a true artifact of the ancient Maya calendar. A rational and academic approach to this topic would be to ask the question: is there evidence we can explore to help us understand what the ancient Maya thought about this date? That's the question I asked and that launched my 2012 work several decades ago.

---

[6] I became aware of the 13-Baktun period-ending through a serendipitous encounter with Frank Waters' 1975 book, *Mexico Mystique*, at age 12 in the summer of 1976 (Jenkins 2007:35-36; 2009:98).

The first part of this book will give a quick primer on Maya cosmology and 2012, with a focus on my findings. Although one could explore a wide spectrum of 2012-related topics, my focus is on the critiques of my work. It is the necessary prelude to the rest of the book, where a narrative of my experiment is provided. That is the necessary foundation of this experiment.

A quick synopsis here (to be explored in more detail in Chapter 2). As a self-educated outsider to official Maya Studies, I did serious research into an edge topic (2012) and by 1998 I produced a well-documented and argued interpretation of what 2012 meant to the ancient Maya. I was communicating with scholars, tried to get my work published in academic venues, but had to settle for publishing through a trade publisher. (The reason why being that academic publishers did not, in the mid-1990s, believe 2012 was a valid topic of rational inquiry.) Various scholars denied, wanly dismissed, or ignored my work. My efforts to educate and communicate with scholars continued through the years. As the year 2012 got closer, scholars started to more aggressively denigrate and maliciously critique my work, with jabs at my personal beliefs and background. I was a "charlatan"; I was supposedly trying to be a "Gnostic New Age prophet," a "Y12er," or I was engaging in astrological pseudoscience and had plagiarized my work from some astrologer (Aveni 2009, Freidel 2009, Freidel & Villaseñor 2009, Hoopes 2011).

In 2006 a Maya inscription mentioning the 2012 date — one that most scholars were not even aware of — came to widespread attention. This forced a more serious consideration of the 2012 date within academia, but the diehard deniers of 2012 were resistant. Reacting to the increasingly absurd marketplace mess, these deniers were adamantly fixated on the idea that 2012 meant nothing to the Maya. This attitude continued in some prominent circles, unabated, until the very end of the year 2012, despite growing evidence to the contrary. Meanwhile, a few professional Maya scholars who were open-minded started to offer interpretations for what 2012 may have meant to the ancient Maya. Lo and behold, the new emerging consensus (among these few scholars) explored and echoed the very same ideas I'd been working out and articulating, on my own, for two decades.

So, scholars who were late comers to the topic of 2012ology (as I coined it), or "the 2012 phenomenon" (as my friend Geoff Stray called it in 2002) were beginning to treat 2012 seriously. Each was finding their way to Room 2012, entering cautiously, only to find me sitting there, offering an orientational hand-out. For many, this was unacceptable. The academic equivalent of barroom bouncers were called in to oust me. To be more specific, a reflex within academia was initiated. And it may be that this reflex was largely unconscious — there were probably no conspiratorial memos handed down (although some backroom collaborations do seem to have happened). It's just that scholars cannot, to save face, allow major discoveries and breakthroughs to have been accomplished, and published, by an outsider to their field, especially while they were on record denying the relevance of the topic for years.

The first "officially acceptable" article dedicated to exploring 2012, written by a degreed scholar and published in a peer-review journal (the "new religions" journal *Nova Religio*), appeared in 2006 by literature professor Robert Sitler. In it, he accurately summarized my new discoveries and distinguished my 1998 book as "by far the best researched of the numerous books that focus on the 2012 date." This was followed by a thesis paper in 2007 (Defesche) which asserted there wasn't much in my 1998 book that was new, and then a lengthy Power-Point style "article" by Van Stone in late 2008 (which ignored the information I provided in lengthy email exchanges with him earlier that year). These three widely different takes on my work underscore what happens when scholars do one of three things: 1) read my work (Sitler); 2) don't read it but simply dismiss it (Defesche); 3) ignore my work even after it's hand delivered and explained to them (Van Stone).

Sitler's positive assessment of my work was embedded within a larger critique of the marketplace appropriation of 2012, which inadvertently influenced modern Maya leaders. In other words, Sitler wasn't concerned with the question of reconstructing what the *ancient* pre-Maya — the creators of the 2012 calendar — thought about 2012. Taking a cue from Geoff Stray's website or his 2005 book, or from my comments in communications we had in 2004 (while he was writing his article), Sitler used the phrase "the 2012 phenomenon" in his title. Later scholars (e.g., Whitesides & Hoopes 2012) incorrectly credited Sitler with both coining and defining this phrase. Hoopes & Whitesides (2014) and Whitesides (2015) have adamantly maintained this stance, despite the facts. You see, the Ivory Tower narrative must keep everything tightly bound within the circle of "real" scholars. Then Maya scholar Anthony Aveni's book of 2009 came out, published with the University Press of Colorado, and the race was on among scholars to line up for their marketplace share of proceeds. Other 2012 books by scholars followed in quick succession (Restall & Solari 2011, Stuart 2011, Van Stone 2010, Sitler 2010).

None of these sources had much to say about what the ancient Maya may have thought about 2012 (with the exception of Sitler's 2010 book which, based on accumulating evidence, weighed toward accepting my work). They were mostly unconcerned with looking at the evidence or with engaging in any kind of risky trail-blazing exploration. Instead, they played it safe. They were addressing modern attitudes to 2012 or making a misleading point (by now a boring trope) about there being no explicit statements about a Maya doomsday in 2012.

In all this, the *modus operandi* for scholars was set, on how to talk about 2012. They would simply critique the "phenomenon" of 2012 as it appeared in the hyped media and the marketplace. This was low hanging fruit, bolstering an emerging safe consensus among scholars. Most of the media marketplace was obsessed with some kind of doomsday in 2012, or sudden spiritual ascension, UFO visitations, Planet X, or crop circles somehow being connected to the Maya calendar. It was easy to point out (as I'd been doing for years) that there was no evidence the Maya thought of 2012 as a doomsday. Furthermore, most of the marketplace productions on 2012 were highly dubious from an academic viewpoint, ranging from underinformed research to creative model-making and philosophizing, to survivalist strategies and outright silliness. My 1992 book *Tzolkin: Visionary Perspectives and Calendar Studies*, critiqued some of this as well as the unfortunate rise of personality-driven spiritual materialism in the 1980s New Age marketplace.[7]

---

[7] See pp 167-168. Reprinted with Borderland Sciences Research Foundation (Garberville, CA) in 1994.

And then there was my 2012 work. It was well argued and presented in my 1998 book *Maya Cosmogenesis 2012* — my ninth book on Maya-related topics since 1989. The book was almost always cited or mentioned by scholarly critics, but was likewise almost always inaccurately portrayed or dismissed through false assertions about its content, false assertions about my approach and methods, or by taking out of context partial quotations (see the Krupp section). For most knee-jerk scholarly critics, the challenge was to force-fit my work into the same denigrating category that all the other 2012 writings must be crammed — an invented modern "mythology" or "Mayanism" consisting of utter nonsense, astrological pseudoscience, prophetic blathering, and poppycock. According to Nicholas Campion, I was part of a dubious "Maya Prophecy Movement" (MPM), but "real" scholars who explicitly bandied about the "2012 prophecy" idea were given a pass (see my exchange with Campion in Appendix 1).

After 2006 there followed many years of false, denigrating, misleading, and malicious *ad hominem* treatments of myself and my work, broadcast all over the Internet, Youtube, Wikipedia, as well as in peer-review journals, academic blogs, and respected university press books and journals. There's no point in taking to task the more slippery manifestations in blogs, slanderous debunking websites,[8] Youtube, or on Wikipedia — although scholarly critics often mentioned or drew from these as proof for my dubious status. (Many presentations by scholars at academic venues were uploaded to Youtube and document the case I will make, so the Internet can be helpful). Rather than waste time sparing with this army of trolls, *my focus will be on university presses, peer-reviewed journals, and a scientific institution (NASA) — all of which have official errata policies in place.* I sought dialogue with the upper echelon of Maya Studies, their academically empowered publishers and legitimizing agency (the AAUP). This is the critical factor in my Experiment, and how it proceeds as a test.

I've selected six primary scholars (Morrison, Aveni, Krupp, Hoopes-Whitesides, and Carlson) who published in peer-reviewed journals or books, or on the government-run and tax-payer supported NASA science website, and have asked them to apply their own professional policies and standards of fact checking and error correction. When that failed, I went to a "higher court", so to speak, in asking their academic publishers to objectively assess the situation. When that failed, I went to the umbrella committee of the organization (the AAUP, the Association of American University Presses) that grants legitimacy to university presses and supervises their behavior. When that failed, well, that's when we grade the test and share the facts and evidence to expose institutionalized corruption. We start to see that the Ivory Tower is more like a house of cards. The behavior of authors, editors, publishers, and their oversight committees all come into play.

My main selections for the Experiment are just a slice of a bigger pie, a much deeper and larger iceberg. So in Appendix 1, I sketch various other perpetrators. But that's not all. I also include, in Appendix 2, online links to additional exchanges and documents. Most importantly, I conceive this book as a readable

---

[8] Such as Johann Normak's "Archaeological Haecceities" blog or Bill Hudson's "2012Hoax" website — both contaminated by malicious cyber-stalker Jim Smith (a.k.a. "Tom Brown").

narrative spiked with some entertaining wit-satire, and have relegated the indispensable primary resources to another appendix (Appendix 3). These files include all of my requests, complaints, and "error files," word for word and unedited, that I sent to the authors and relevant policy committees and press directors. These files will be frequently cited throughout the narrative of this book, lest readers think I'm just making this all up or otherwise doubt what actually transpired in email exchanges. This is like having the insider's view of what really happened, which indicts in no uncertain terms the corruption of trusted press directors, officials, and scholars.

Since Appendix 3 grew to a book-sized archive of over 100,000 words, I've decided to do something creative. I have made this archive of facts and evidence available, for free, online. There is an easily accessed website dedicated to it, with links to the main file and the sub categories (see p. 158). I will summarize the contents of the appendices and provide links to the free eBook archive. This is about academic transparency, honesty, and review, and those are the raw documents that must be made available for discerning scrutiny and assessment.

This narrative could be extremely lengthy, exhaustive, and tedious. And maybe it is, even in this reduced form (i.e., minus the appendices). It is a big story, one that I've found necessary to document carefully in order to preserve facts, truth, and accuracy on a topic that should have been revolutionary for Maya Studies. The complete version of this story runs to over 1800 pages of documentation, printed out and collected, in three volumes, preserved in a black suitcase in my office. For the purpose of this presentation, I've exercised extreme editing skills to pare it down to a minimized narrative. The inclusion of the lengthy appendices would make for a very beefy book, but that can be detached as a free online eBook resource and can be considered the easy-to-reference documentation, should readers want to check and confirm my summaries in the body of the narrative.

The five chapters of this book can be considered to be what an informed and non-biased investigative journalist might produce, if they studied and understood all the events and documentation. These days, I can't expect journalists to be much interested in 2012. Sadly, when they had the chance to produce some good investigative journalism, most of them torched it and exploited the silliest, most superficial farces associated with it. Now, it's supposedly an expired topic. But hey, journalists, why do we still ponder and write about the JFK assassination? Or WWII? Or the death of Marilyn Monroe? Because we always want to better understand things that have happened in the past, especially things that were, at the time, fraught with hysterical emotions, deceptions, fear, stupidity, and misrepresentation.

That being said, one may ask: to what end my book? Am I seeking vindication? No, that would imply I am out to get people. Rather, I'm very aware that 2012 was completely abused, distorted, and misplayed, not only in the mainstream media, but — more disappointingly — within the hallowed halls of Academe. I've continued to do presentations and interviews up to the present (October 2015) and it seems to me that, for the most part, most people don't understand what went down.[9] So, the

---

[9] Radio interviews and presentations 2013-2015: see http://www.TheCenterfor2012Studies.com.

purpose of releasing this book, and my motivation in writing it, is to clearly assemble the facts and evidence which, without my effort, would end up lost in the sands of time or tossed in a dumpster after I'm planted six feet under. I address this to future historians, and dedicate it to all the historians, past and present, whose work would have been easier if the letters and documents of past generations weren't carelessly destroyed or thrown away.

Thinking people who were not particularly interested in Maya traditions or 2012 might in any case be interested in seeing what happens when a dedicated and informed outsider, like myself, makes a breakthrough contribution to a field of study. Well, we know what happens, courtesy of Thomas Kuhn's *Structure of Scientific Revolutions* — the person making the breakthrough gets mitigated while their work is eventually adopted, as if it was known all along. They get kicked into the gutter while their insights are adopted by trusted professionals. My Experiment provides a textbook example of that process. However, unlike previous episodes like this throughout the history of science, here the truth is exposed for all to see in close proximity to the events, rather than being reconstructed decades or centuries later by concerned historians — long after the victims and perpetrators have died.

So, yes, I suggest that a 2012 redux is in order, especially from the vantage point presented in this book. It's not a book for the UFO, prophecy, Planet X, and crop circle aficionados who populated the 2012 scene five or ten years ago. It is, as my work has always been, for thinking and intelligent readers. Certainly our world has other things to worry about. Then again, I hear all manner of human interest stories on NPR, Story Core, Tedx Talks, On the Media, and other popular media venues — stories that, to me, sometimes seem much more trivial and obscure than our late *cause célèbre*, our much abused 2012 topic.

There's no reason why 2012 shouldn't be revisited and intelligently chewed on in the public discourse, much like other "events" that took place in the past. World War I, the JFK assassination, 9-11, the 15th-century rebellion of the Welsh prince Owain Glendower — books are still being written about all these topics. And there is still disagreement about most of them, so I shouldn't expect 2012 to ever get resolved in the public consensus. But academic consensus on a topic generally does happen, although it takes time, and up to now the negative consensus on 2012 in academia is totally contradicted by the evidence. Consensus should not trump evidence in academia, but it often does, and this indicates that science has become functionally broken. How does that happen? Well, that happens when scholars break it; when they, their publishers, and their supervising checks-and-balances committees violate their own principles and policies.

The data, facts, and evidence I've provided here should be factored into any academic or media treatment of what 2012 meant to the creators of the Long Count, and who figured it out first. That's the name of the game in academia, and probably explains all the anti-JMJ mitigation tactics. (According to 2012 critic Anthony Aveni, "JMJ" is acceptable code for me, John Major Jenkins, but my name and books must otherwise never be mentioned.)[10] Getting the 2012 story right

---

[10] Aveni speaking to his colleagues at the 2008 *Society for American Archaeology* conference.

should matter to modern cultural observers and social commentators, as well as to future historians, and it matters to me.

Bullies and bullshitters exist everywhere in contemporary culture, and I believe it's important to call them out and hand them a report card on their behavior. They are the ones who prevent progress from happening. It takes courage, persistence, and patience to shine the light on such darkness and maliciousness, entrenched as it is within institutions lacking conscience, giving priority to legal advice and pressured by the ego politics of authoritative and self-serving establishment scholars. I can't expect anyone else to do it, so here it is.

◎ ◎ ◎ ◎ ◎

Chapter 2. Maya Cosmology and 2012

A study of "Maya cosmology" includes a host of related disciplines: astronomy, cultural beliefs, mythology, calendars, literature, and religion. In practice the most frequently referenced areas are astronomy and mythology. It is well known that in the Maya worldview these two areas work together. They are interrelated. The Creation Mythology of the Maya *Popol Vuh*, or Hero Twin Myth, is filled with astronomical features involving Venus, the Dark Rift in the Milky Way, and the cross formed by the Milky Way and the ecliptic. In the mid-1990s a rational investigation of 2012 led me into a reconstruction of ancient Maya cosmology that broke new ground, synthesized disparate threads, built upon previous scholarship, and offered several new discoveries. I reconstructed why the ancient creators of the Long Count intentionally placed the cycle ending on December 21, 2012, and what they thought about it. My two-part reconstruction, sometimes called the "2012 alignment theory," integrates astronomy and ideology — that is, astronomy with the mythological and religious beliefs connected to that astronomy.

Any rational investigator examining the possible importance of the 2012 cycle-ending date within Maya cosmology encounters two facts: it falls on a solstice, and it falls in an era of rare astronomical alignment within the precession of the equinoxes. This rare alignment has been called the galactic alignment, or solstice-galaxy alignment, and is real astronomy. It has been noted in *Hamlet's Mill* by science historians Santillana and von Dechend (1969), in D. & T. McKenna (1975), and has been calculated by astronomers (Patrick Wallace in 1999 and Meeus 1997). The treatments of this era-2012 alignment by academic critics has at best been vague and dismissive, the critiques being predicated on a distinct avoidance of the evidence and arguments I've brought to bear on the question, in many books, presentations, and articles stretching over twenty years.

My journey and evolving thoughts on 2012 are well documented in my published books, going back to 1989. Prior to that, the date was in my thoughts from readings in popular and academic literature beginning around 1985. Prior to that, as I recounted in my book *The 2012 Story* (2009) and my article in *The Mystery of 2012* anthology (2007), the 2012 concept was seeded in my young mind, at age 12, by a serendipitous encounter with Frank Waters' book, *Mexico Mystique*. That encounter was providential and memorable, occurring on the Bicentennial celebration of July 4, 1976. It was like a whisper, nothing more and nothing less, inviting me into my future intellectual preoccupations. Of course, I didn't do much with it at the time,[11] but I remembered it ten years later when I was reading Frank Waters' *Masked Gods* and saw *Mexico Mystique* in a used bookstore, its dark blue dust-jacket with the Olmec head and that cool and suggestive word, *mystique*, triggering the old memory.

When I first traveled to Mexico and Central America in 1986, the year 2012 wasn't so much on my mind. Rather, I wanted to visit temple sites, make friends among the modern Maya, and learn about their 260-day sacred calendar. The 2012 date was a fact of the system, and it simply became an accepted part of the calendar tradition I was studying. It didn't become a focus until 1993. Still, at several junctures I did scratch my head about it, and these early thoughts can be found in *Journey to the Mayan Underworld* (1989), *Tzolkin* (1992/1994), and *Jaloj-Kexoj and Phi-64* (1994). Early on I realized that Frank Waters' book was problematic in certain central assumptions, but I felt he was correct in several of his insights. The 2012 "end date" (actually, *cycle*-ending date) was, somehow, about astronomy — probably involving the precession of the equinoxes. And it was part of a World Age doctrine that was expressed in the Maya Creation mythology. That was the basic orientation, which has proven to be accurate. Waters' thoughts on 2012 were updated in a little-known essay he wrote in the late 1980s, published in a posthumous anthology called *Pure Waters* (2002), edited by his wife, Barbara. In 2014 I reviewed it and a link to my essay can be found at *The Center for 2012 Studies* website.

The purpose of this chapter is to report my published interpretations of 2012, documenting how my research identified evidence for my interpretations, primarily after I focused on 2012 in earnest in 1993. I've concisely identified and described my 2012 reconstruction many times, as for example in the following letter:

In a nutshell, in my study of the pre-Classic site of Izapa (the culture and site that many Maya scholars believe was involved in the formulation of the Long Count calendar), my study of ballgame and Creation Myth symbolism, and king-making rites, I've argued that **the creators of the Long Count intended the 2012 period-ending date to target a rare astronomical alignment within the cycle of the precession of the equinoxes, and saw this alignment as signaling (not definitively *causing*) the need for deity sacrifice in order to facilitate worldrenewal.** (Letter to Griffith Observatory Director Ed Krupp)

That's it, honed for simple presentation. You won't find that exact verbatim phrasing in my earliest writings on 2012, and it's unrealistic to expect to. The very process of writing and re-writing, debating and discussing, hones the language used to express such newly reconstructed conceptions. But the idea of sacrifice as the

---

[11] It did seem to trigger a science fiction story I wrote shortly thereafter, which used the date "November 11, 2013" as the date of a nuclear war, from which the protagonist planned to flee via cryogenic freezing into the future. It was sci-fi; I was twelve years old and was trying my legs as a storyteller.

necessary prelude to world-renewal was there, elaborated from the evidence at Izapa, in my 1996 monograph *Izapa Cosmos*, and in my 1998 book *Maya Cosmogenesis 2012*. In addition, the idea of a like-in-kind parallel between the Creation dates in 3114 BC and 2012 AD (an idea used as a key in John B. Carlson's work of 2010, as well as in Carl Callaway's work of 2011), is expressed in my 1995 book *The Center of Mayan Time* and my 1995 online review of Stuart & Houston (1994), with a nascent suggestion of it in my 1994 book *Jaloj Kexoj and Phi-64*. It's an idea that makes total sense, and can be argued from the evidence, although 2012 debunkers in academia often treated it (especially when I expressed it) as being unfounded and ludicrous.

In 2002, my book *Galactic Alignment* concisely summarized my prior work, and we see my two-part reconstruction more concisely expressed. This was still many years before scholars entered Room 2012. In a 2010 lecture John B. Carlson stated that the last time he spoke about 2012 was in a 1993 talk (see Carlson section below), but he's never produced a written typescript, recording, or summary of his lecture. His talk was timed to the Katun commencement of April 5, 1993, and thus a mention of 2012 — when the new Katun would close — would have been an obvious calendrical inference. That may indeed be all he spoke about. In comparison, I wrote and published my thoughts on the 12.19.0.0.0 Katun commencement in my 1992 book *Tzolkin*. I pointed out that the upcoming Katun shift was oddly coordinated with both a full moon and a Venus morningstar appearance. It would be interesting if Carlson noted these curious correspondences, but he's refused to respond to my various emails and queries. We'll never know for sure unless there is reliable, published, dated documentation on what his talk entailed.

Even more recent publications of mine, including my 2009 book *The 2012 Story* and my SAA lecture of April 2010, preceded any publications by Maya scholars on what 2012 may have meant to the ancient Maya. They later started echoing my previously published ideas, and despite two decades of efforts at dialogue and communication with them, they have largely neglected acknowledging my prior work. So, let's look at some choice quotes from *The 2012 Story*. They are not hard to find. On pages 2-3 of the book, in my Introduction:

My 1998 book *Maya Cosmogenesis 2012* broke new ground on identifying why 2012 was important to the ancient Maya, offering a new reconstruction of ancient Maya thought. Key questions were posed: When and where did the early Maya devise the calendar that gives us the cycle ending in 2012? Why did they place this cycle ending on December 21, 2012, and how did they think about? These questions led me to discoveries and conclusions that integrated the domains of astronomy, mythology, prophecy, and spiritual teachings.

I found that a rare astronomical alignment culminates in the years leading up to 2012, when the position of the solstice sun will be aligned with the Milky Way galaxy. This solstice-galaxy alignment is a rare occurrence, happening only once every 26,000 years. It can be called a "galactic alignment" and was perceived by ancient astronomers as a shifting of the position of the sun, on the solstice, in relation to background features such as stars, constellations, and the Milky Way.

Based on evidence in Maya traditions and key archaeological sites, it became overwhelmingly apparent to me that the future convergence of sun and galaxy was calculated, with good accuracy, by the ancient Maya and the cycle ending date in 2012 was chosen to target it. Without going into any further questions and complexities, this situation means the ancient Maya had astronomical abilities at least on par with their contemporaries in other parts of the world including Greece, India, Babylonia, and Egypt.

Importantly, I noticed that the astronomical features involved in the galactic alignment were key players in Maya cosmology and Creation mythology. These connections were not free floating opinions based on imagined associations that had no real relevance for the ancient Maya. In fact, the evidence was there in the academic literature itself. I was merely stitching all the pieces together. The solstice sun, the Milky Way, and a curious feature that lies along the Milky Way called the dark rift, were utilized in the sacred ballgame, king making rites, the calendar systems, and the Hero Twin Creation myth. These real connections anchored the galactic alignment firmly within known Maya concepts and traditions. In my studies I quickly focused my attention on the early Maya site called Izapa, which scholars suspected as being involved in the formulation of the Long Count calendar. By 1994 the results of this approach had revealed Izapa as a critically important place for understanding how the Maya thought about the galactic alignment in era-2012. Furthermore, the astronomy was woven together with spiritual teachings, conveyed as mythological dynamics in the Creation myth on Izapa's many pictographic monuments.

Those "mythological dynamics in the Creation myth" express the relationship between deity sacrifice and worldrenewal in era-2012. One can look up "sacrifice and renewal" in the Index and find discussions in various places in the book. Scholars critiquing my work after 2009 could have embraced my actual words, expressing my position, but very few of them did. In fact, there are zero quotations of my work — that's right, you will not find one complete sentence quoted from my 2009 book in the later critiques offered by the scholar-critics. In fact, the same is true for my 1998 book *Maya Cosmogenesis 2012*. They'll paraphrase, always misleadingly, or take clips out of context. To me, that is not good scholarship — unless your intention is to mislead and malign, to craft false and baseless denunciations.

We just surveyed my 2009 book; now let's go back to the beginning. I always refer to my 1994 *Mountain Astrologer* article as the first published announcement and articulation of my early discoveries — the astronomical key to 2012. It was written in May of 1994, and contains some early but clear definitions, connecting the controversial "galactic alignment" astronomy to concepts in Maya cosmology and the Creation Myth. After all, *this* connection to known Maya concepts was my unprecedented breakthrough work that legitimized 2012 as a valid topical of rational, scholarly, investigation:

The dates on which the sun conjuncts the "Sacred Tree" are thus very important. These dates will change with precession. Schele doesn't pursue this line of reasoning, however, and doesn't even mention that these dates might be significant. If we go back to 755 A.D., we find that the sun conjuncts the Sacred Tree on December 3rd. I should point out here that the Milky Way is a wide band, and perhaps a 10-day range of dates should be considered. To start with, however, I use the exact center of the Milky Way band that one finds on star charts, known as the "Galactic Equator" (not to be confused with Galactic Center). Where the Galactic Equator crosses the ecliptic in Sagittarius just happens to be where the dark rift in the Milky Way begins. This is a dark bifurcation in the Milky Way caused by interstellar dust clouds. To observers on earth, it appears as a dark road which begins near the ecliptic and stretches along the Milky Way up towards Polaris. The Maya today are quite aware of this feature; the Quiché Maya call it xibalba be (the "road to xibalba") and the Chorti Maya call it the "camino de Santiago". In Dennis Tedlock's translation of the Popol Vuh, we find that the ancient Maya called it the "Black Road". The Hero Twins Hunahpu and Xbalanque must journey down this road to battle the Lords of Xibalba. (D. Tedlock 1985:334, 358). Furthermore, what Schele has identified as the Sacred Tree was known to the ancient Quiché simply as "Crossroads" (Jenkins 1994)

Notice how I built on the work of previous scholars, who nevertheless had not put the pieces of the 2012 puzzle together. Of the several dozen references to, and discussions of, my work by scholarly critics, nowhere can be found a description of my work that accurately conveys my perspective on the evidence and data, as expressed in the quote above, from my very first article on the topic, written in May of 1994. This article has been publicly and freely available online since 1996 (on Terence McKenna's website and my own website). I continue in the paragraph following the above quote:

This celestial feature [the Crossroads] was not marginal in ancient Mayan thought and is still recognized even today. In terms of how this feature was mythologized, it seems that when a planet, the sun, or the moon entered the dark cleft of the Milky Way in Sagittarius (which happens to be the exact center of the Milky Way, the Galactic Equator), entrance to the underworld road was possible, which could then take the journeyer up to the Heart of Sky. Shamanic vision rites were probably involved in this scenario. In the Yucatan, underground caves were ritual places used by shaman to journey to the underworld. Schele explains that "Mayan mythology identifies the Road to Xibalba as going through a cave" (*Forest of Kings*, 209). Here we have a metaphorical reference to the "dark rift" in the Milky Way by way of its terrestrial counterpart, a syncretism between earth and sky which is characteristic of Mayan thinking. Above all, what is becoming apparent from the corpus of Mayan Creation Myths is that creation seems to have taken place at a celestial crossroads - the crossing point of ecliptic and Milky Way (Jenkins 1994).

In this same article I make a clear distinction between the astronomical alignment process under discussion and any "astrological" interpretations that may be secondarily applied to it:

> It is critical to understand that the winter solstice sun rarely conjuncts the Sacred Tree. In fact, this is an event that has been coming to resonance very slowly over thousands and thousands of years. What this might mean astrologically, how this might effect the "energy weather" on earth, must be treated as a separate topic. But I should at least mention in passing that this celestial convergence appears to parallel the accelerating pace of human civilization (Jenkins 1994).

It is reasonable to point out, as I did, the empirically unusual "alignment" quality of modern times in coordination with the Long Count period-ending date of the Maya calendar. This circumstance can be an opening to further considerations, as Barbara MacLeod and Mark Van Stone alluded to, *eighteen years after my article was published*, in their award-winning 2012 essay in *Zeitschrift für Anomalistik*.

My 1994 article does not explicitly explore how the Maya interpreted the era-2012 alignment. My main concern was to document the astronomical alignment process, and I had not yet delved deeply into the other question. However, there are inklings in this early article. The concluding paragraph reveals my focus on the Maya World Age doctrine, involving the end of one World Age as an opening channel up the central conduit of the Sacred Tree, into "the Heart of Heaven."

> This essay is not contrived upon sketchy evidence. It basically rests upon two facts: 1) the well known end date of the 13-baktun cycle of the Mayan Long Count, which is December 21st, 2012 A.D. and 2) the astronomical situation on that day. Based upon these two facts alone, the creators of the Long Count knew about and calculated the rate of precession over 2300 years ago. I can conceive of no other conclusion. To explain this away as "coincidence" would only obscure the issue. For early Mesoamerican skywatchers, the slow approach of the winter solstice sun to the Sacred Tree was seen as a critical process, the culmination of which was surely worthy of being called 13.0.0.0.0, the end of a World Age. The channel would then be open through the winter solstice doorway, up the Sacred Tree, the xibalba be, to the center of the churning heavens, the Heart of Sky.

Notice I write the "end of a World Age," not the end of *the world*. Such a distinction is also present in my 1989 book *Journey to the Mayan Underworld*. Even in the early stages of my 2012 work I didn't assume the "final end" of the doomsday meme. Why didn't I? Because of the evidence I was finding as to how the Maya thought about it. Note that at this early stage I followed Schele on her "Black Transformer" and "Pole Star = Heart of Sky" conceptions, which I soon re-thought in favor of my "Three Cosmic Centers" thesis. The visionary ascension of a Maya king-shaman into the cosmic center was symbolic of his accession to rulership, conferring upon him that status. King accessions were conceived by the Maya as the birth of a new being (with the concomitant death of an old one) and utilized the upturned frog-mouth glyph, which means "to be born." So, embedded in even this early articulation of the ideological belief associated with the alignment, we see an indirect allusion to the "rebirth" concept. My later research honed this as the understanding that "deity sacrifice is necessary for worldrenewal," confirmed by the dialectic between Seven Macaw and One Hunahpu in the Izapan ballcourt (Jenkins 1996, 1998).

I wanted to show a comparison between my early published writing in 1994 and my 2009 book, *The 2012 Story*, which was published in hardback and paperback by Tarcher/Penguin Books. The extensive promotion for that book included my interviews on the Hollywood Red Carpet for the 2012 disaster movie, in which I expressly emphasized that 2012 was not about doomsday.[12] There were numerous other appearances at the time, including Fox News (the Hannity Show), an ABC Nightline program, and (in 2007) a feature piece in the *New York Times*. How many authors get this kind of exposure and publicity? Within a few months of my book's release I presented my work at the academic SAA conference. I have presented and published in both popular and academic venues. So, my presence in the discussion could not be dismissed (well, yes, it *could* be dismissed, which is largely what happened). The vast majority of scholarly critique of 2012 and my work occurred *after* 2009 and therefore my book, *The 2012 Story*, could have been referenced for my position. But none of the scholars did so, in any kind of specific, accurate, or relevant way, even up through Krupp's 2014 article (he references only my 1998 book, which he has used to take my words out of context).

The two-part reconstruction I have offered, involving astronomy and an ideology of renewal that requires sacrifice, is presented and argued from the evidence in my 1998 book *Maya Cosmogenesis 2012*. The core of the evidence is found in the symbolism of Maya Creation mythology, king-making rituals, and the ballgame. All of these facets come together and are expressed as a unified paradigm in the ballcourt at Izapa. The previously unrecognized piece that unites the astronomy and ideology is the astronomical orientation of the Izapa ballcourt, which I was the first to calculate and publish (1996, 1998). Several years after this, Anthony Aveni and Horst Hartung published their field data (in 2000) which affirmed my previous work. It was published in an obscure journal and I wasn't aware of it until 2009. It would have been helpful, but Aveni didn't share it with me during our 2008 email exchanges. In fact, in his 2009 book Aveni stated the ballcourt orientation incorrectly, some 48 degrees in error. He and his publisher (and then the AAUP) denied this error, which I document here in this book (see Chapter 3b).

My point is that Part IV of my 1998 book contains the Izapa ballcourt work, which is rooted in an interdisciplinary analysis of archaeoastronomy, archaeology, iconography, calendrics, ballgame symbolism, the surrounding topography of "environmental determinants"[13] (Isbell 1982), iconography, and the Hero Twin Creation Myth. In a brief nutshell, my two-part interpretation is based in the orientation of the ballcourt combined with the dialectic between Stela 60 and the Throne carvings. From this is derived the idea that "deity sacrifice is necessary for world-renewal," that the Creation Day events in 3114 BC and 2012 AD are like-in-

---

[12] Among others, a *Time Magazine* interview resulted, which is online (see the Krupp exchanges).
[13] See excerpt here: http://alignment2012.com/mc-dialectic.html.

kind bookends during which similar Creation Myth dynamics could be expected to unfold, and that an alignment of the December solstice sun with the Dark Rift in the Milky Way was expected (or calculated) to occur at the end of the World Age — symbolic of the rebirth of One Hunahpu and the victory of the ballgame (the gameball going into the goalring). A shift away from the older polar paradigm of the Olmec thus occurred, and a new "cosmic center" orientation was adopted. All of these ideas are found later expressed in the 2012 interpretations of Carlson, Callaway, Coggins, MacLeod, and Grofe, *after* 2009.

As I survey my published articles and books on 2012, I'd say my article for the anthology *The Mystery of 2012* provides a clear presentation of my ideas, geared to a popular readership. It was written in August of 2006, and published in 2007. I worked with Sounds True to contextualize various writers on 2012. I suggested a section that would include writers who were working to reconstruct what the ancient Maya thought about 2012. However, the problem here is that I would be the only person in that category. I was living in Boulder when Tami Simon founded Sounds True around 1985. Their office and recording studio used to be located above the Crystal Market on Pearl St, and it was fun to visit the tape library, browse the sale shelf, and chat with people who came and went. Sounds True recorded conferences and it was a great way to be informed about talks that you couldn't attend in person. The 1980s passed, Sounds True expanded. It made sense that they would contact me to contribute to their planned 2012 anthology — the first of its kind in the marketplace.

By mid-2006, it was time for another clear presentation of my work in brief form. The previous year I had written a manuscript called *Approaching 2012*. Inspired by the fireside-chat style of Peter Kingsley's *In the Dark Places of Wisdom*, my book just laid it all out in a conversational tone. I pitched it around to publishers, and there was interest, but the deals offered were appalling. The negotiations began with "no advance and a 10% royalty." That was the face of publishing at the time, despite being a proven and bankable author on a hot topic. Meanwhile, other books in the marketplace were divorcing 2012 from needing to have much reference to Maya traditions — I think here of Daniel Pinchbeck's book *2012: The Return of Quetzalcoatl* (2006), which apart from his Intro didn't even mention 2012 until after page 100. Or, for another example, "2012" books were increasingly about alarmist doomsday threats (Joseph's *Apocalypse 2012*). Within this publishing feeding frenzy, a thoughtful book like my *Approaching 2012*, that delved into the lost cosmology of Izapa — the origins of the 2012 paradigm among the Maya — went to the bottom of the slush pile. So, that book was never published. However, the style of presentation was still with me when I wrote my article for the Sounds True anthology.

The title of my article was "The Origin of the 2012 Revelation." I used that term "revelation" because what happened at Izapa was akin to a radically new religious movement, rooted in new cosmological discoveries. Furthermore, I've frequently noted the fact that shamanic tools of vision were used at Izapa, and thus a visionary and spiritual *revealing* of new insights and perspectives no doubt contributed to the formulation of the Long Count cosmology. It occurs to me now that my title allowed undiscerning critics to seize upon a false notion about my work. Anthony

Aveni stated in his 2009 book that "Jenkins' ideas have not been well received among mainstream Maya scholars, who place little stock in subjective analogies and knowledge acquired through revelation" (Aveni 2009:23-24). He seems to be suggesting that my work was granted to me as a revelation, a sort of fickle dream, rather than as the result of many years of study and theorizing from the evidence.

In an essay published in 2012, Kevin Whitesides & John Hoopes adopted a similar view (see the Hoopes section), in which they asserted that my interpretation of the cosmology at Izapa was derived from a subjective archetypal assumption, which I believe doesn't require scholarly analysis or disputation. Although they cite my 1998 book to support this bizarre hallucination about my methods, such a reading of my work cannot possibly be informed by actually reading my work. One can understand Aveni's misunderstanding of my work if he merely glanced at the title of my article, and assumed that I was suggesting that the "Izapa revelation" was something that struck me and resulted in my ideas. No, that isn't what I discuss within the article, nor did I present my 2012 ideas as coming to me in this way. For example, in my article I wrote:

> In 1998, when I completed my book *Maya Cosmogenesis 2012*, I was confident that my findings were supported by the evidence — that I had put a key concept on the table that made sense of so many disconnected threads in how official scholars interpreted Maya traditions. The very fact that the 13 Baktun end-date fell on a solstice date highlighted some kind of intention behind it, but scholars dismissed this as coincidence. I was bemused and baffled at how any progress could take place in academic circles when this kind of attitude runs the show. Yet, I was following the trails laid out by the scholars themselves. Michael Coe, for example, had said that "the priority of Izapa in the very important adoption of the Long Count is quite clear cut." (Jenkins 2007:45)

I also discussed Maya spiritual traditions, a topic which I suspect scholars who are steeped in scientific materialism just smear over and inject their own personal judgments into.

But what of the well-known role of intuition and insight, "hunches," in scholarly process? There can be a process of integrating and synthesizing information from disparate disciplines, which is part of the thoughtful head-scratching that can result in a new realization — the proverbial light bulb going on over ones head. Sometimes this is called having a "hunch," and then following through with assessing evidence. To some degree this has been part of my process, and new discoveries have also emerged from synthesizing information from my wide readings in various academic disciplines. (I have frequently pointed to the bibliography for my 1998 book, which documents the sources I've read and integrated into my work: http://Alignment2012.com/bibbb.htm; some 95% of them are high-level academic studies, essays, and monographs.) Making new connections from the data can result in breakthrough discoveries and a new picture emerging, which is what happened in my work. I suspect that most academic brains don't work in this way — they are good at connecting the data-dots but not at drawing the picture, let alone explaining the picture to others.

A core synthesis in my work is how I showed that the astronomical features involved in the galactic alignment of era-2012 are front and center in the Maya Creation Mythology. The Dark Rift in the Milky Way, and the Crossroads of the Milky Way and the ecliptic, are the conceptual targets for the galactic alignment calculation. I particularly emphasized the role of the Dark Rift, connecting to the 2012 alignment astronomy. In the Wikipedia entry for "The 2012 Phenomenon," this breakthrough idea has been incorrectly credited to Munro Edmonson's 1988 book, *The Book of Year*.

John Hoopes has been a primary contributor to crafting this entry. We see this, for example, in Hoopes's own comments to Mark Van Stone's FAMSI essay on 2012, which was released in late 2008. Hoopes wrote, and I paraphrase his words precisely, that he was involved, on an ongoing basis, in a "behind the scenes dissemination" of info on 2012 and related topics. This was accomplished, wrote Hoopes, through his contributions to Wikipedia entries such as: the 2012 phenomenon, Mayanism, Votan, Hunab Ku, and Charles Étienne Brasseur de Bourbourg. He says he has been "trying to take advantage" of Wikipedia in order to help the popular public understand 2012 and "its associated mythology." Hoopes's comment was posted on the FAMSI website around 2009 and remains there to this day: http://www.famsi.org/research/vanstone/2012/comments.html.

Hoopes's Wiki efforts at crafting his interpretation of nefarious influences on "the 2012 mythology" was also recommended by John B. Carlson, when he announced the IAU "2012" anthology on the Aztlan e-list in mid-2011. I've observed how Wikipedia has served the 2012 debunkers as a dissemination venue for false and incorrect information. It's somewhat tricky because in addition to completely false assertions, as in the example given above of crediting Edmonson with my work, doubt can be injected into various ideas through clever phrasings and loaded lingo. This was rife on Wikipedia, in all the 2012-related websites. The Izapa website, for example, for years stated there are no ballcourts at Izapa. This helps the polemical denunciation of my work. In addition, in mid-2010 my own name/author entry on Wikipedia was hijacked by an anti-2012 debunker, an Anthony Aveni fan named Jim Smith. It's very difficult to police the disinformation on Wikipedia. To this day, the "2012 Phenomenon" entry does not accurately state who coined and first used that term (see pp. 90-97 in the Whitesides & Hoopes article from 2012, and my response to it in Jenkins 2014a: www.update2012.com/Jenkins-Zeitschrift-fur-Anomalistik-1-2014.pdf). The injection of doubt into the narrative, by debunker skeptics like Hoopes, is the same strategy employed by scientists who are in the pocket of tobacco lobbyists and climate deniers, as explored in the book *Merchants of Doubt* (Oreski and Conway).

These problems in the portrayal of 2012 and my work stem from two shortcomings: mis-readings of my work because of a careless, superficial assessment; 2) maliciously intentional distortions and omissions. This chapter provides some excerpts from my work, and some contextualizing explanations that will address the flawed readings and assertions that have resulted from these substandard treatments. Apart from the explicit statements in my work which critics should be able to cognitively process, slightly more nuanced explanations of difficult concepts are well-written and presented in my work. Degree-holding

24

scholars should have no problem understanding and honoring them. However, they almost always do have problems, because they are adamantly fixated on mitigation and denigration. They are, basically, unwilling *to learn* what my work is about. Their agenda is to mitigate by rejecting or ignoring facts. To maintain that agenda they must violate the policies and standards of science.

There is so much that I could present, by way of illustration, regarding the nature of my work compared to the unconscionable and inaccurate treatment of it by scholarly critics. However, the purpose of this chapter is to sketch my work and to emphasize that it has grown out of a progressive though seemingly aborted milieu within Maya Studies. I refer here to the work of Linda Schele and a few other scholars in the 1980s and early 1990s. The direction toward which things were unfolding, after the publication of Schele & Freidel's *Maya Cosmos* book in 1993, were initially hopeful and positive. In an article in *Archaeology* magazine (July/August 1993), Maya scholar Peter Mathews said that the new connections being made between Maya myth and cosmology "open up a whole new world of discovery. We stand on the threshold of something truly new."

This was followed by slap downs courtesy of Anthony Aveni. In his review of *Maya Cosmos*, Aveni invoked his favorite trope, that the authors were indulging "emotional revelations" — the same critique that he leveled against me in his 2009 book. Consequently, Maya Studies swung back towards a regressive suspicion of seeing astronomy within Maya mythology and symbolism. Aveni's review was published in 1996, the very year I reached out to Aveni and other scholars, seeking dialogue on my work which was developing into my book *Maya Cosmogenesis 2012*. We see here the parting of the ways. I was willing to follow wherever the new evidence led, while Aveni's camp fell back into the smug security of *the already known* — incomplete models that resisted new evidence.

But even earlier than this the brakes were being hit, regarding where Schele's work could have led. Schele herself was resistant to ideas connected to 2012. I wrote her a letter in May of 1994 and sketched my understanding of the precessional alignment that would culminate in the years around 2012. My angle of approach was to point out to her that, throughout the 700 years of the Maya Classic Period, the alignment of the sun with the Dark Rift/Crossroads would progressively occur on different dates in the Tropical Year, approaching alignment with *the solstice* in era-2012. Thus, we might look for dates in the inscriptions that corresponded to the sun's alignment with the Dark Rift/Crossroads, slowly shifting through the Tropical Year during the Classic Period. My letter was brief:

Dear Linda Schele,   May 16th, 1994

I'd like to share some of my thoughts that may contribute to your work with Mayan epigraphy and cosmo-conception. I'll try to be brief. Since you have identified the crossing zone of the ecliptic and Milky Way as a significant Mayan concept — none other than the Sacred Tree — I'd like to point out some related considerations. First, during the Mayan heyday, where was the crossing point in relation to the winter solstice? In other words, there must be a specific yearly date when the sun comes around to conjunct this point. Right now, this date is approximately the winter solstice. Some 1500 or 2000 years ago, the date would have been some 20 to 27 degrees out. If the precession of the equinoxes was noticed by the Maya when the Long Count was inaugurated circa 300 B.C., and they noticed that the precession was causing the "Sacred Tree" to slowly approach the winter solstice, a forward calculation to 12.21.2012 may have been made to calculate when synchronization would occur. ... 1) consider the dates when the sun conjuncts the Sacred Tree at various historic times in Mayan History; this changes with time and may explain the Creation Day [12.21.2012] ... 2) We might want to consider precession, a forward Long Count calculation at the Long Count's inauguration, and the unique astronomical situation revolving around 2012 (winter solstice sun conjuncting the Sacred Tree) as the how and why of end point 12.21.2012. ... If time permits, it would be great to hear your thoughts on this idea. If not, thank you very much for all your breakthroughs. [unedited and verbatim excerpt of a one-page letter]

She did not respond to me, although she did respond to a letter from Milo Rae Gardner who inquired about 2012 slightly earlier that year. I know this because Gardner also sent me a similar query. Schele's response to Gardner's 2012 questions was published on the nascent University of Texas Mesoamerica e-list in April 1996, and was combined with a reaction to the popular, and deeply flawed, 2012 book called *The Maya Prophecies* (1995). I was privy to all this as it unfolded, and was somewhat surprised, because I met Schele in March of 1995. She acknowledged receiving my letter of the previous year. My comments to her in person were packaged with the importance of Izapa, so she merely directed me to speak with her grad student, Julia Guernsey, who was studying Izapa.

My approach of looking for inscriptional dates that placed the sun at the Dark Rift/Crossroads was mentioned in my 1998 book. A breakthrough of sorts occurred in early 2000, when I realized that the dedication date of Copan Stela C was one of these dates. Furthermore, the inscription and iconography on the stela supported the galactic alignment astronomy. This was the kind of separate support, from another angle or field, that helped to validate my thesis. Intention overrules coincidence when this happens. At least, it should, but only if scholars are not overly fixated to their dogma of coincidence.

One scholar, Michael Grofe, found my Copan Stela C work worthy of citing. It was, after all, published in two places: 1) The IMS *Explorer* (December 2000) and 2) in my 2002 book *Galactic Alignment*. In fact, in February of 2009 my approach triggered Grofe's discovery of the galactic alignment birthday astronomy of Tortuguero's Lord Jaguar. Unfortunately, in later publications Grofe framed my statements inaccurately in a way that — and this was quite difficult to experience — set my work off to the side as jumping to a conclusion that was essentially different from his approach. But if you read my actual words in my 2000 essay, as well as in my 2002 book, one can see that I was not jumping to a conclusion, but was suggesting that we could explore this kind of astronomy as it might manifest in a variety of Maya traditions, just as Grofe was suggesting.[14] In my original version (from 2000) I did not even mention 2012.

---

[14] See http://thecenterfor2012studies.com/Copan-Stela-C-Jenkins2014.pdf

Sadly, the problem was compounded by his revised SAA 2010 article of 2015, in which my Copan Stela C discovery was stated, but I was not cited for it or mentioned. Compared to an earlier version of his article, several mentions of me and citations to my writings were removed. I suspected that, because Grofe's piece was published in Carlson's *Archaeoastronomy Journal* (Vol. 25, released in March 2015 but pre-dated to a "2012-2013" publication date), a prudent redaction was made by the author or requested by the editor. Instead of redacting, Grofe might have *added to* the revised version of his piece, which after all deals with Copan astronomy, a mention of his discovery confirmed during *The Great Return* conference (December 18-23, 2012). But to this day that discovery remains unpublished, and my own elaboration of the implications of Grofe's discovery, in a completed 3000-word essay, languishes unseen.

Without giving away Grofe's discovery, the title of my unpublished essay conveys the implications: "Deciphering Copan Stela C: A Key to Classic Maya Calculations of the Sidereal Year and the Tropical Year with Special Reference to the 13-Baktun Cycle Era Base and the Sun's Alignment with the Milky Way/Ecliptic Crossroads (as occurs in AD 2012)." And here's a direct quote from my unpublished essay:

Now, suddenly in late 2012, we have a breakthrough in understanding this monument. At the end of this essay, I could generally summarize that "Stela C indicates a sophisticated astronomical knowledge of the Classic Maya at Copan," but this would severely understate the case. The renewed effort to understand this monument comes with an added two-part twist: 1) the key to this new, unprecedented breakthrough in understanding Stela C occurred *on the morning of December 21, 2012 at the site of Copan*, and 2) the resulting newly reconstructed astronomical framework confirms that the Maya were interested in, and were able to calculate, "galactic alignments" such as the one that features as the centerpiece of my much-criticized reconstruction work regarding what the ancient Maya thought about the 13-Baktun period ending on December 21, 2012. (Jenkins 2013, n.d.)

Here are a few other choice quotes from my various other articles and books:

To clarify, hopefully once and for all: I am not saying the Maya predicted the solstice-dark rift alignment with exact precision, and my theory does not require exact precision. I am not saying that the alignment happens only once on the solstice of 2012 (it happens on winter solstices within a range of 2012). I am not saying that the alignment causes pole flips, solar flares, or anything necessarily. I am not saying that the ancient Maya believed the alignment signals the end of time, the end of their calendar, or the end of the world. All I am saying is that the alignment of the solstice sun with the dark rift in the Milky Way is demonstrably the empirical phenomenon in nature that the ancient creators of the Long Count were intending the 13-bak'tun period ending in 2012 to mark, indicate, or target.

This core idea in my pioneering work is now receiving new support from the information contained on Tortuguero Monument 6. The challenge, as with any data, is how thoroughly it is understood. A superficial treatment of Tortuguero Monument 6 "doesn't tell us much," as David Stuart said, but a systematic and thorough reading of the text, with sensitivity to its astronomical and numerological themes, tells us that my "galactic alignment theory" was barking up the right tree some fifteen years ago. The evidence points to the role played by the solstice sun's alignment with the dark rift in Maya cosmo-conception, kingship, creation mythos, and building dedications. And it reinforces the notion that 2012 was conceived as a cosmological renewal, a calendrical and mythological creation event inextricably interwoven with the recognition by the ancient Maya that 13.0.0.0.0 fell on a solstice and on that future day the sun would be aligned with the Crossroads of the Milky Way and the ecliptic at the southern terminus of the dark rift in the Milky Way. That is the crux of the reconstruction I first published in 1994 and elaborated in my 1998 book *Maya Cosmogenesis 2012*. (Jenkins 2011b)

That quote was from my chapter in the Gelfer anthology, which Michael Coe wrote the introduction for. Here is the Abstract (first paragraph) and my Summary of my presentation at the *Society for American Archaeology* (April 15, 2010):

Abstract
First, my "2012 alignment" hypothesis will be clearly defined. I will present evidence in the Classic Period inscriptions of Tikal, Copán, and Quirigua, with a special focus on Monument 6 from Tortuguero, for the use of the dark rift in the Milky Way as a reference point for planetary, lunar, and solar alignments. Using a new method of schematically diagramming a complex hieroglyphic inscription, an analysis of a repeating astronomical theme in the thirteen dates recorded on Monument 6 strongly suggests an awareness of the sun's future alignment with the dark rift in the Milky Way on the solstice of 2012 AD, the 13-Baktun period ending recorded in the right flange of that monument. The methodology acknowledges and incorporates textual references that are not exclusively phonetic, namely astronomy and astronumerology, enabling a fuller reading of the intended meaning.

Summary
This has been a very brief treatment of a topic that deserves a more detailed presentation. Of the 13 dates on the Tortuguero monument, six involve alignments of the sun, Jupiter, and a lunar eclipse with the dark rift/Crossroads, with possibly five additional dates of significance to the dark rift. Based upon the pattern of astronomical references on the 2012 monument from Tortuguero, it's likely that the people of Tortuguero intentionally used an awareness of the sun's future alignment, on a solstice, with the dark rift in the political rhetoric of a 7th-century king. Furthermore, the pre-existing calendrical structure of the Long Count, having been developed centuries prior to Tortuguero, requires that the knowledge of the 2012 alignment of the solstice sun and the dark rift/Crossroads was embedded into the Long Count at its very inception, over 2,000 years ago.

The evidence presented here argues that the dark rift/Crossroads was utilized as a reference point by the Classic Period Maya in a veritable symphony of alignments involving the sun, the moon, planets, eclipses, and the solstice position of the sun.

Overall, it appears to be involved in rituals and ideation relating to sacrifice, rebirth, transformation, period endings, building dedications, and king making. This Classic Period evidence invites a more serious and factually accurate assessment of my earlier archaeoastronomical reconstruction work on precession and dark-rift astronomy at Izapa.

My SAA presentation was a written report, read out loud. Later in 2010 the paper was reviewed and critiqued in the public debate set up by the scholars at the Maya Exploration Center. The resulting transcript of that discussion/debate runs to over 92,000 words and was posted online under the Maya Exploration Center's research files. I had to balance several simultaneous conversations, and upon conclusion of the debate I was invited to summarize the experience:

Eventually, Maya Studies may have to acknowledge my work as pioneering and unprecedented, and disregard the inaccurate conflations of me with other areas of the 2012 mess. In any case this will all move beyond any vindication of my work. Even if that should be forthcoming, I'm sure it will be mitigated by caveats, continuing misconceptions despite my best efforts, and I'll be relegated to unsubstantiated rumor and innuendo placed in footnotes. The larger concern of progress in Maya Studies is the necessary, and long overdue, integration of astronomy and epigraphy. Maya Studies has suffered from a pendulum swing between these two areas with concomitant related polarizing between history and mythology, etic versus emic approaches, and so on. We are, I believe, awaiting the final phase in Hegel's thesis-antithesis-synthesis process. Let's not feel we need to take sides in this perennial dualist debate, but let's integrate the equally valid concerns of both sides. Perhaps this discussion served to catalyze a step forward in this much needed integration, moving things away from one-sided views and toward a truer reflection of how the Maya themselves actually viewed their world.

As we acknowledge and accept the full complexity of Maya thought, a more complex approach is necessary, one that does not take safe harbor in cookie-cutting new proposals through narrow filters but, instead, acknowledges the full spectrum of data and evidence that can and should factor into any honest theory, model, or reconstruction. (Jenkins, December 19, 2010, in Jenkins 2011)

Here's an excerpt from my article in the Institute of Maya Studies *Explorer* (Vol. 39, Issue 12, December 2010):

But why? Why use a personal connection to a great period ending in the Long Count as a strategy for amplifying ones political power and divine status as king? Simple: that's what Maya kings did. That's what Janaab' Pakal did in exploiting the 20th-Bak'tun period ending in 4772 CE; that's what K'ak Tiliw of Quirigua did in relating himself to the astronomical three-hearthstone event (in 3114 BCE); that's what 18 Rabbit of Copan did in connecting himself to deep time mythic rituals and a future 10th-Bak'tun period ending. What Lord Jaguar did with the 13th-Bak'tun period ending in 2012 was simply par for the course in the Maya kingship playbook: they were interested in exploiting real or asserted connections with the calendrically-defined Creation Mythos. Namely, using period endings, preferably big ones. Thus, 12/21/2012 was of interest to Lord Jaguar because of the astronomical parallel his birthday had to it. What this means, however, is that he or at least his political rhetoricians must have been aware that the sun, on the solstice day of 12/21/2012, would be aligned with the center of the Crossroads formed by the Milky Way and the ecliptic, at the southern terminus of the Dark Rift in the Milky Way. (Jenkins 2010)

My article for the anthology *You Are Still Being Lied To* (2009) concisely summarized my work, and framed it in the context of other contributions to Maya Studies made by outsiders:

When I studied the Maya Creation Mythology, their ballgame's symbolism, and the carved monuments of Izapa (the place that invented the Long Count), I found previously unrecognized evidence that the ancient Maya became aware of this future galactic alignment some 2,100 years ago. Furthermore, an entire galactic cosmology was embedded in these Maya institutions, involving astronomy as well as prophecy and spiritual teachings. So, the keys to my work are the galactic alignment, the Creation Myth's symbolization of the galactic alignment, and the underappreciated site of Izapa. Although my work has proceeded rationally with careful documentation, no one else asked the right questions, and so my findings are unprecedented.

Although this is the unavoidable approach to 2012 that any rational investigator would take, scholars have been barred from the path by the limiting dictates of their professed Coincidentalism. While often refusing to actually examine my work, scholars have unfairly tended to see me as belonging to the irrelevant arena of New Age speculation. Thomas Kuhn wrote that most major breakthroughs in an evolving field of study are made by self-taught outsiders— precisely because they are not in bed with the biases and assumptions that keep progress from happening. Goodman, Teeple, Knorosov, Proskouriakoff—Maya Studies is in fact filled with these independent pioneers. Recently, new discoveries regarding the astronomical knowledge of the Maya have been made by scholars, notably by epigrapher Barbara MacLeod, effectively mitigating previous criticisms of my work (Jenkins 2009).

My 1997 presentation at the IMS is now fully transcribed, based upon a video taken of the entire event (it's at *The Center for 2012 Studies* website). In it, we see my early articulation of the sacrifice and renewal doctrine applied to 2012, as well as an explicit statement that I didn't think 2012 was about doomsday. The essay on Izapa that I worked on, and proposed for publication in Carlson's *Archaeoastronomy Journal* (in 1999) and in Nicholas Campion's *Cosmos and*

Culture Journal (in 2000) was ultimately published online in early 2001. It is titled "Izapan Cosmos: A brief survey of Izapan iconography and astronomy in the Group F ballcourt." The result illustrates what might have been introduced into the official conversation many years prior to scholars taking 2012 seriously. Some excerpts from this piece (online: www.alignment2012.com/Izapa.html):

> What follows is a brief exploration of Izapan iconography and orientations in the light of horizon astronomy. This material was first published in a lengthy monograph (*Izapa Cosmos*, 1996) and later incorporated into my book *Maya Cosmogenesis 2012* (1998). Since these publications I have felt that Izapan iconography and astronomy needed more attention, and perhaps supportive illustrations and diagrams would be helpful to illustrate what we find. In this brief recapitulation, I will focus on the monuments of Group F, for they clearly suggest something profound and unrecognized about the Izapan awareness of astronomy. ...
>
> Stela 25 also contains a recognizable *Popol Vuh* episode, in which Hunahpu's arm is torn off by Seven Macaw. Since Seven Macaw is identified with the Big Dipper of the polar region, I suggested (1996, 1998) that the "fall" of Seven Macaw involved the demise of an old cosmological system centered upon the polar region. The shamanistic concern with knowing where the center of the sky is located is central to understanding this "cosmological shift." [Note: these are ideas reiterated by Coggins in a 2015 publication.] The shift, after Seven Macaw was done away with, was to an opposite orientation, as revealed in the diagram above. The dialectic sets the head of the alligator in opposition to Seven Macaw. This "alligator-head" is the location of another cosmological "center of the sky"—it is the location of the center of our Milky Way galaxy. (Note: I am omitting arguments and citations that can be found exhaustively documented in my book *Maya Cosmogenesis 2012*). Generally, this monument—as well as many others from Izapa; e.g., Stela 11—indicate an interest in the Milky Way, the dark-rift in the Milky Way (the mouth is the dark-rift), and the Big Dipper. ...
>
> One thing is interesting to consider: Given the integrative continuity of my interpretation, based upon a complicated set of interweaving monuments, sculptures, calendrics, and alignments, is it likely that Group F does not have anything to do with the precessional convergence of Milky Way and solstice sun? (Jenkins 1999-2001)

Finally, there are many concise summaries of my approach and work in my 2009 book *The 2012 Story*. My 2002 book *Galactic Alignment* also contains a section which concisely summarizes the evidence underlying my findings, as explored in detail in my earlier book *Maya Cosmogenesis 2012* (1998).

I want to emphasize that my work has not been operating completely outside of academic venues and publications. I have bridged academic and popular opportunities. Critics can, and will, dismiss the status of anything they find irrelevant or distasteful, even if presented in peer-reviewed journals or in university programs (as has my work, including classes taught at the fully accredited Naropa University in 1999). Many of my articles and presentations have occurred in cite-able publications and academic venues. I know this because the "official" and "real" Maya scholars have been cited for their contributions to some of the very same publications and venues that my work has appeared in. Some cite-able examples:

- My 2010 SAA presentation, with a public peer-review sponsored by Maya scholars and published at the *Maya Exploration Center* in 2011.
- My 2014 *Zeitschrift fur Anomalistik* review-essay
- My 2014 article in *Clavis Journal*
- My many articles in the Institute of Maya Studies *Explorer*, 1997-2017.
- My 2011 chapter in Gelfer's 2012 anthology, with Intro by Michael Coe
- My classes and presentations at the accredited Naropa University, in 1999
- My two presentations at the Institute of Maya Studies, in 1997 and 2011
- Other invited presentations, classes, and panels at various institutes, universities, and colleges.

My interviews and breakthrough work have appeared in *The New York Times*, the *U.S. News and World Report*, the *Atlantic*, *Newsweek*, the U.K. *Guardian* and *Time Magazine* online, CNN Science online, *Archaeology Magazine*, *Sky & Telescope*, and in various documentaries. The double standard of scholars is revealed in that they have readily approved and cited the 2012 books of their colleagues which were either self-published (as in Van Stone's 2012 book) or published with trade publishers (as in the case of Stuart's, Sitler's, and Restall & Solari's books on 2012), yet my 2009 book with Tarcher/Penguin is panned because it was not "academically published."[15]

Van Stone exclusively grants validity to these books, including his own, and the lack of rational processing going on among scholars is exposed in his belief that only five books on 2012 have been published by legitimate, degreed scholars. Well, in fact, popular writers Carl Calleman and José Argüelles both held PhDs when writing their books. They are deeply flawed books, and so each author and book should be judged on its own merits, which is exactly how I have treated, reviewed, and critiqued the 2012 offerings. Calleman's and Argüelles' books are error-riddled and conceptually problematic, to the say least. As are the books of all the "real" scholars, including himself, that Van Stone perceives as being on some other level simply because they were written by degree holders. This is a lazy, factually flawed, elitist, and undiscerning attitude — the same style of attitude that has plagued the treatment of 2012 and my work by "real" scholars.

The devils are in the details, so now it is time that we look at what scholars have actually said. The following chapter shares what I call my "experiment," in which I ask scholars, their scientific employers, and their peer-reviewed publishers to confirm and correct various errors they have stated and published. Will scientists

---

[15] Almost all of the scholarly articles and books on 2012 were published after my 2009 book *The 2012 Story* was released, yet it was never cited. Scholars would default, in their laziness and inattentiveness, to my earlier books such as *Tzolkin* (1992/1994), *Maya Cosmogenesis 2012* (1998) or *Galactic Alignment* (2002).

behave like scientists? Will academic university-press publishers abide by their policies and principles? Let's find out.

Chapter 3. The Experiment: Asking Science to do Its Job

The provocative title of this chapter should be qualified. I am not, of course, addressing my Experiment to all of scholarly publishing. As I alluded to in Chapter 1, the Experiment addresses how the field of Maya Studies has treated the topic of 2012. As such, the question involves a specific case within Maya Studies. This specific illustration does, however, indicate how any field (scholarly publishing in general) can get violated if the conditions are right.

In this chapter I share five different efforts to ask scholars, their academic publishers, or the scientific institution they work for, to acknowledge and correct factual errors, as required according to their own stated policies. Each example reveals and illustrates how scholars violate science, often in collusion with their scientific and academic publishers. The treatment of Hoopes's "Mayanism" construct and his unsupported slanders published in the *Archaeoastronomy Journal* requires a focus apart from his essay co-written with Whitesides. The Whitesides-Hoopes debacle brought the issue into sharp focus and unfolded in 2013-2014, resulting in the 2014 peer-reviewed publication of my review of their flawed essay. This effort was wrapping up as I started to strategize my approach to the other four examples.

These four got officially underway in early 2015, although they all have deeper roots. It was like simultaneously engaging four, long drawn-out chess games. In three of the examples (Aveni, Carlson, and Krupp) there was a previous history of contact and exchanges, going back in all three cases roughly twenty years. The remaining example involves NASA scientist David Morrison, who I'd never communicated with before and I was only vaguely aware of his efforts and how he was intimately connected with some of the other anti-2012 scholars. Some pointed research made the connections more clear, and clearly indicts him as an under-informed, malicious, and biased debunker. I'll begin this chapter with him.

The Life and Times of Dr. Doom

David Morrison has many accolades and impressive academic credits. You'd think he would be quite capable of performing fact-based and rational assessments of another person's work. But maybe he forgot. I don't think it's unreasonable for me to expect that my critics state factual things about my work and background. However, in Morrison's case his statements are 100% false, completely inverted as to the true nature of my 2012 work. Let's first get to know Davey M a little bit and then we'll take a look at what he's said about me, over and over again in his presentations at academic venues between 2009 and 2013, as well as on the NASA website.

Morrison specializes in the field of astrobiology. His work's focus for decades has been on tracking asteroids and comet impacts on earth. There's a lot of space junk flying around up there, and we know from studying geology that sometimes the earth gets smacked. Serious strikes happen at long intervals, but we better be on the look out, because if a chunk of rock hit us sometime soon, it would be devastating for life on earth. Luckily, the geniuses at NASA have been looking out for us. If they could just get some more tax payer money to track space junk trajectories, we could anticipate possible impacts and shoot them out of the sky with laser beams. Zap! Pow! Boom!

Throughout the 1980s and into the 90s David Morrison repeatedly spoke to representatives and funding agencies in Washington DC, to get funding for these tracking projects. He played the card of impending doom so much that he was given the nickname Dr Doom. He learned very well that striking fear into the heart of the public could trigger a larger concern in policy circles such that the coffers would open to fund multi-million dollar projects. Dr Doom seemed to like his moniker; he even recalled it with pride during a talk he gave in early 2009. In March of 1990 he published, with Clark R. Chapman of the *Planetary Science Institute*, an article for *Sky & Telescope* magazine called "Target Earth: It *Will* Happen."[16] Yes, the emphasis on *Will*. The cover depicted a monstrous asteroid slamming into the earth. As you might suspect, the alarmist fear-mongering title could easily scare people into believing the end was near. A few years later, the article was cited and described as "frightening" in a scholarly book about comet impacts.

How was it, then, that Dr Doom emerged as an anti-doomsday, anti-2012 crusader? Well, it's hard to know how this occurred. I have some suspicion that he was influenced by anti-2012 activist, amateur astronomer, and "death to the infidels" imperial Inquisitor Bill Hudson, who had started a website called 2012Hoax.com. In 2009 he became upset at how the rising tide of 2012 doomsday rhetoric — driven by the impending Hollywood movie — was scaring little children. On his blog he sent out a call to eradicate all the noxious "2012 proponents" and offered to pay lawyers to file suits against them. He wrote: "I want them to go down. Hard." Within weeks he launched his 2012Hoax domain, and many jumped aboard to take down those noxious 2012 proponents. He set it up as a member-defined information outlet, supposedly like Wikipedia. But in practice it became a billboard for malicious axe-grinding debunker trolls, like Southern Baptist math teacher Jim Smith. By mid-2010 there was a bio profile on me and misleading critiques of my work, mostly courtesy of Smith. Clearly, I was presented as a "2012 proponent" meaning, in the lingo of the 2012Hoax website, a doomsday guy. It took me months of effort in early 2012 to educate Bill Hudson on how this was a completely false and slanderous portrayal of my position. Eventually I was shifted to another category, "2012 author."

Morrison must have been following the evolving 2012Hoax website, as he cites it several times on his NASA "Ask an Astrobiologist" page, which fielded questions

---

[16] http://starfoundation.nousphere.net/wp-content/uploads/2015/01/Target-Earth-it-Will-Happen1.pdf.

about 2012 for over four years up to the end of year 2012. However, Morrison's first assertions about me occurred over a year before Hudson's website was launched, at the very beginning of his Ask an Astrobiologist phase. There, someone had asked him where the 2012 doomsday association began. Morrison replied that, in the Maya calendar arena it was me, John Major Jenkins, who began in the 1990s writing that the Maya predicted doomsday in 2012. The statements appear three weeks apart:

"The claims about the Mayan calendar predicting catastrophe in 2012 began with John Major Jenkins (a self-styled "independent researcher") in the early 1990s." (David Morrison on his NASA page, February 17, 2009).

"In the case of the 2012 hoax we have a confluence of pseudoscience websites that are trying to cash in on the growing fear of a cosmic disaster in 2012… The roots of the 2012 hoax lie in misinterpretations of the Sumer/Babylonian mythology by Zecharia Sitchin and of the Mayan calendar by John Major Jenkins" (David Morrison on his NASA page, March 9, 2009).

I was not aware of these statements until much later, although I'd run across Morrison's blog several times over the years. I found it to be one of many attempts by scholars to address the ridiculous doomsday-2012 meme. One passage I saw revealed that Morrison did not understand what the galactic alignment was. Morrison's anti-2012 comments showed that he considered 2012 to be synonymous with doomsday — a very undiscerning and superficial position. I never saw any of his presentations until 2014, when they started getting posted to Youtube. There may have been a few posted earlier, but I had too much on my plate to be tracking everything being said, all over the internet, and it didn't seem that Morrison was targeting me specifically. In fact, from what I could see, I would agree with him taking to task all the alarmist doomsday pimps, scaring little children. I'd dealt with my fair share of emails from kids and their parents, and gave an anti-doomsday presentation on Y2K and 2012 to Denver high school students as long ago as 1999. However, what I didn't realize was that Morrison had fingered me as the main source of the 2012 Maya calendar doomsday mess.

In doing some investigative research, I soon discovered five or six presentations on Youtube and elsewhere, given by Morrison at conferences sponsored under the academic auspices of organizations like Griffith Observatory, NASA, and SETI. I was fairly aghast at his totally false characterization of me. He liked to use a slide that contained a picture of me, lifted off my website, next to the cover of my 2009 book *The 2012 Story*. He would present me in a sequence of 2012 doomsday proponents including Pleiadean channeler Nancy Leider and Zechariah Sitchin. His slide read: "Maya Apocalypse" and he'd say ridiculous things about my background and beliefs about 2012. Somewhat later in his presentation, he'd say that all these 2012 people were responsible for at least one suicide and for scaring little children. Should such utterly false character assassinations remain online unchallenged? Well, no, according to NASA's truth and honesty Communications Policy, applicable to all NASA employees including David Morrison.

I decided to approach this in two phases, so as not to overly alarm the perpetrator and his employer. First, I would inform the director of the Communications Office about the two statements on the NASA website. This would be closest to home, and within reach for an easy fix. I sent my first email, filing my complaint to their office, on February 5, 2015. No response. Weeks pass, I sent a follow-up email. No response. I make several phone calls, leave messages with a secretary, until finally I get the ear of the director. He will look into it. Weeks later I'm informed that one of the statements was deleted and a change was made to the other one. But the change did not effect the false assertion made by Morrison. All in all it took over three months for the two statements to be correctly addressed. Curiously, they were not to be explicitly acknowledged or changed to a correct statement about my work; rather, they were simply to be deleted.

This was a minor victory, accomplished after numerous emails and phone calls stretching over three months. David Morrison himself was alerted to the process, but I didn't, earlier in the process, feel it was prudent to communicate directly with him. I felt the correction needed to be addressed through the NASA Communications Policy office. But now the door was open. I succeeded in getting a brief response from him — all too brief. Now I would go to the next phase of the chess game. I would ask him to apply the same correction to all his presentations online. I would make it very easy — all I needed from him was a written acknowledgement that his statements about me and my work were wrong, factually incorrect. Scholars are suppose to acknowledge and correct errors when proven wrong. I provided all the proofs in my second filed complaint.

And so the next phase of evasions and delays began. I sent several emails directly to Morrison, cc'ing his Ames Research Center supervisor and the Communications Office in Washington, DC. No response. So, as of early September — seven months after I initiated this experiment — we find Harvard graduate David Morrison failing to behave like a responsible, ethical scientist. Under pressure from his employer, he deleted the two comments on his blog, but he would not admit that they were totally false statements. That's what I need to effectively correct the record in relation to his presentations that are still posted on the Internet. But he won't do it; he prefers to violate the values and principles of science. He is therefore not a practicing scientist, but a disreputable non-functioning scientist. His violations help to break science.

This component of the Experiment was partially successful, but it revealed a deeper dysfunction of a broken system — it may be forced to grudgingly delete false information, but it will resist admitting that errors were made. The arrogance and superiority complex of academic egos is at work here, and we also see this problem in the incredible resistance of Anthony Aveni, and his university press publisher, to admit to even the simplest of errors (let alone correct them). With that example, explored next, I had to bring it to the "higher court," so to speak, of the AAUP that confers academic validity on their university press members, and which assesses violations of academic standards and stated member policies.

Again, in my Experiment I was simply asking these scholarly institutions to enforce their own stated policies. David Morrison, rather than jumping at the

36

chance to practice science, and apply an intellectually honest policy of acknowledging errors, instead did the minimal amount necessary and ultimately evaded acknowledging that any error was made. The game could be pursued forever, and I could take it to a "higher court" of NASA's office of investigations (as climate scientist James Hansen did), but the message is clear: the online errors must be protected and maintained, because they serve the purpose of falsely mitigating my name and my work. With the example of Anthony Aveni, this mitigation reflex is even more likely — whether it be consciously employed or just an unconscious reflex against outsiders — because he is more closely connected to Maya Studies than Morrison.

For the details of my efforts with David Morrison, please read in Appendix 3 my complaint filed with NASA's Communication Office on February 5, 2015, as well as my emails to Morrison, which were cc'd to my contacts at the NASA Communications Office (see online links on p. 158).

Memo from the Bully Pulpit

Unlike the David Morrison/NASA story, my exchanges with Aveni go back many years, to 1996. There are several episodes with Aveni prior to my formal complaint sent to the University Press of Colorado and the Association of American University Presses (AAUP) in 2015, which I'll sketch before we sink our teeth into the events of 2015:

1. Exchange in 1996 and invitation to receive my book, 1997
2. Aztlan discussion w/ Lloyd Anderson re Aveni, precession, and my new book, in June 1999
3. The NYTimes piece (2007), followed by the Milbrath exchanges in the IMS *Explorer* in which I responded to Aveni's critique.
4. The MacLeod-Grofe-Jenkins dialogue of mid-2008, with revealing comments on Aveni, and my simultaneous email exchange with Aveni (April 2008).
5. The Tulane event, documented in my 2009 book.
6. My review of Aveni's article promoting his book (November 2009).
7. Snide and false statements made in Aveni's presentations (e.g., in early 2012)[17]
8. The exchange with Aveni about my Benfer piece, in late 2013.

In our brief exchange of 1996, Aveni disagreed with my belief that the Maya had a doctrine of World Ages. This was presented in my 1995 article "Maya Creation: The Stellar Frame and World Ages," which I sent him. Aveni's position against the World Age concept is odd but persistent through the years. It necessarily ties into a consideration of the precession of the equinoxes, as precession underlies the doctrine of World Ages. As such, in 1999 I engaged a discussion with scholars on the Aztlan e-list, regarding the precessional basis of my work on the era-2012

---

[17] This presentation is linked prominently to his personal web page, and is posted on the Colgate University Youtube channel: https://www.youtube.com/watch?v=qItCYBhZui8. I am among those he labels as being "prophets of Y12."

astronomical alignment (the so-called "galactic alignment" or "solstice-galaxy alignment"). I quoted Aveni's positive position on precessional knowledge among the ancient Maya, found in his 1980 book *Skywatchers*: "Ancient astronomers easily could detect the long-term precessional motion . . . Through myth and legend the earliest skywatchers transmitted their consciousness of the passage of the vernal equinox along the zodiac from constellation to constellation" (1980:103). Aztlan member Lloyd Anderson cordially took the lead on engaging my ideas, and contacted Aveni to get his current views. Aveni, now aware that my ideas were being considered on Aztlan, decided to waffle in the other direction and told Lloyd:

> While I cannot specifically document it, I think [the pre-Columbian Americans] probably noticed that rise/ set positions & heliacal rise times changed because this is quite noticeable over, say, a century; but that's a long way from conceiving of precession in the way we understand it, which doesn't come until well after the Greeks (for a discussion see Evans, J. HISTORY & PRACTICE OF ANC. ASTRONOMY). -AAveni

The operative framework of reference for Aveni is revealed in his statement: "but that's a long way from conceiving of precession in the way we understand it." In addition, his statement to Lloyd was a revised position, compared to what he stated earlier in his 1980 book. As an indication of how Aveni's new comment was tailored on the fly to cast aspersions on the favorable consideration of my work, we see that in the revised 2001 edition of *Skywatchers*, he did not alter his previous statement. So, he is able to work it both ways, stating one thing in one place and the opposite elsewhere. This practice is a running joke among critics of academic writing — that "careful" and "safe" scholarship basically involves covering your ass by saying opposite things in different places, so in the event you are critiqued for one position you can call attention to the other. Your rebuttal proceeds: "Actually, what I said was ____" and you can give the impression that your critic is clueless.

Aveni's periodic lambasting of precession and the World Age doctrine is found throughout the years. Along with this is his insistence on the need to measure precession with the shifting rise-times and rise-azimuths of stars, a horizon-based methodology. These rise positions and timings shift with precession, but at different rates depending on whether the star under consideration is near the ecliptic, or far away from it. This approach to tracking precession reveals a Western astronomy bias (the "way we understand it"), and ignores numerous other methods that the Maya could have been employing. In other words, it is *only one* of several methods for tracking precession, and Aveni dismisses the others as being illegitimate, based purely on his bias.

Apart from my own work of the 1990s, alternate proposals on how the Maya may have tracked the precession of the equinoxes, by MacLeod and Grofe, were in place by 2008. They were presented at the Austin conference in early March (MacLeod) and the SAA conference in Vancouver in late March (Grofe). During 2008 I was in an ongoing trialogue via email with MacLeod and Grofe. In it, MacLeod brought up my 1999 Aztlan exchange with Lloyd Anderson and Aveni. She wondered if

Aveni committed to his revised opinion about precession in the 2001 revised edition of his book. Grofe quickly confirmed that he hadn't, and contributed detailed quotes. We all noted how Aveni requires Maya astronomy to be force-fitted into his Western astronomy bias, regarding how precession might have been tracked by the Maya.

For example, MacLeod wrote (July 9, 2008): "Aveni assumes that we credit the Maya with an accurate estimate of the precession cycle based on their supposed observations of 'azimuth shifts'. This is in fact *not* the method we attribute to the Maya ... I do feel that Aveni is blinded by a need to prove that the Maya conceived of precession in the way that we do." MacLeod was spot on in her observation. Ironically, however, this contradicts what his own 2008 SAA presentation was supposedly arguing for:

Aveni, Anthony (Colgate University)
[79] *How Did the Maya See and Interpret Sky Phenomena?*
Creating social order through the perceived template of cosmic order is common in cultures throughout the world. Our safest assumption here is that the way the Maya interpreted what we take to be universally observable phenomena (e.g., the first appearance of morning star, the phases of the moon) is different from our own. I believe this is a factor often neglected in the study of Maya creation stories, largely because of our own lack of familiarity with the sky. This presentation will illustrate how such information illuminated the Maya understanding of space, time, and the process of creation. (SAA conference guide, online, emphasis added)

It seems Aveni says one thing but does another. MacLeod continues: "Furthermore, he has such a knee-jerk response to the notion that they could have deliberately placed the 2012 end date on the winter solstice that he cannot now accept any possibility of evidence for precessional observations in spite of his prior statements."

For his part, Grofe noted (in referring to an email exchange he had with Aveni): "It is very misleading for Aveni to state that he has studied our material 'quite thoroughly.' I have repeatedly asked him for his feedback and criticism of our work, and by his two, short comments below, I can see that he has not thoroughly studied it by any means."

Grofe also revealed, in his first-hand account of two separate instances in 2007 and 2008, Aveni's "vociferous" attempts to dictate a negative judgment of my work to his colleagues, forbidding them from mentioning me by name:

I agree with Barb that Aveni has a clear agenda here of knocking down *Hamlet's Mill* and anything to do with precession and 2012. On both occasions when I presented at the SAA conference with Aveni as the moderator, I have personally heard him vociferously speak out against these things, and against John Major Jenkins' work (he insists on referring to John as "he who will be referred to only as JMJ" — admittedly so as not to draw attention to potential buyers of his work — but Aveni makes such a stink that I hope this has only worked in John's favor!). Now I know how John must feel, as Aveni did not even acknowledge my presence or my work when I presented next to him on the same panel in April. (email of July 9, 2008)

This was all unknown to me at the time. Independently, and unaware that the 2008 SAA conference had just happened in Vancouver, I decided it was time to contact Aveni and thus unfolded an email exchange (between April 23rd and April 30, 2008, which was one month after the SAA conference in Vancouver): http://alignment2012.com/Aveniexchanges2008.html.

One thing to note about this exchange is that Aveni never volunteered his own calculation of the Izapa ballcourt orientation (which would affirm my earlier, independent, calculation and publication of it), despite my insistent request for his comments on my ballcourt alignment work. Again, I was unaware of his 2000 publication at this time. I suppose I sensed his resistance to discussing it, and thus tried to pry some comment out of him, as if intuiting that there was something there, unspoken. And indeed there was. At this point, Aveni refused to divulge his affirming fieldwork at Izapa, which he published *two years after* my 1998 book. Eventually, at Tulane, he did tell me about it (in February 2009), and I immediately looked it up at the Tulane library (after deciphering his boggled recollection of the anthology's title). I then included a mention of it in my 2009 book, which I was then completing.

Such generous last-minute updating of my book was not paralleled by Aveni in his own book, which was released the same day as my own (October 15, 2009). Of course, he may have wrapped up his book before the Tulane conference, but in any case I doubt he would have made any effort to incorporate any corrections or new support for my ideas. In his crafting of an article to promote his book, he used sections from his book that critiqued my work. As such, his interest seemed to lie in making sure his reading public would falsely conclude that my work was flawed. However, his critiques in his book were themselves deeply flawed — so much so that it's a wonder they passed through the peer-review and fact-checking process of his academic publisher, the University Press of Colorado.

That's correct: Aveni's book is filled with errors, specifically about my personal beliefs, about archaeoastronomical alignments at Izapa, and about the precession of the equinoxes — critical to having an accurate and honest basis for judging my work. This incredible circumstance will be the focus of my narrative, below.

When I first read Aveni's book (in October 2009) I was aghast at his errors about my work, but at the time the more egregious problem with his book was his mistaken reading of Grofe's work. I immediately noted some problems, contacted Grofe, and he assessed the issues, found additional errors, and contacted Aveni. Consequently, Aveni was forced to acknowledge three factual errors, and promised he would address them in a future publication. This didn't happen until some six years later, with the release of the corrected edition of Aveni's book (the eBook version), in May 2015.[18]

---

[18] Grofe corrected Aveni in peer-review publications of 2011 and 2012, as I also did in my peer-reviewed chapter in Gelfer (2011).

My review of Aveni's article and book was posted to my website in late 2009. Aveni had plenty of cheerleading colleagues who reviewed his book — Marta Barber at the IMS and John Schwaller on FAMSI, for example. None of them observed the many obvious errors, or found any of his loose assertions and hostile lampooning objectionable, including his use of religious bigotry as a basis for critiquing my work. A young scholar named Kevin Whitesides reviewed Aveni's book on Amazon, and took him to task for his loose assertions and unsupported assumptions. Whitesides was then attacked, on Amazon, by an Aveni fan — anti-2012 activist Jim Smith, who was behind the creation of screeds against me on the 2012Hoax website, as well as on Youtube. In 2010 he also, as previously mentioned, polluted my Wikipedia name-entry page with misinformation — a situation which took weeks to clear up, with the Wiki moderators finally taking a look and quickly determining that he was a malicious troll, not abiding by Wiki standards. See: www.update2012.com/Jim-Smith-Tom-Brown.pdf

Up through 2012 Aveni gave presentations in which he repeated his false and misleading assertions about me and my work. He would use guilt-by-association techniques, such as asserting that I was "a student" or "follower" of José Argüelles.[19] I was largely unaware of these presentations at the time, but today many of them are posted to Youtube and the official video channels of colleges, museums, and universities.

In late 2013 I was still willing to try to engage dialogue with Aveni, and asked him for his assessment of my Benfer piece. Given the recent debacle that had unfolded, in which my much-worked-over chapter was prevented from being included in Benfer's anthology, I thought I'd try to determine who the anonymous "peer-reviewer" was, who leveled ultimatums at Benfer and the anticipated university press publisher (the University Press of Florida), should they include my chapter. I thought Aveni was a possible candidate, so I leadingly asked him for his comments on my chapter. He agreed to take a look. He did respond briefly several weeks later, in a way that makes me disinclined to think he was directly involved in the Benfer debacle (it may have been Dearborn and/or Ruggles).

In any case, what Aveni did say about my piece further illustrates his odd way of not seeing what is actually there. He claimed I had not dealt with the possibility that the astronomical alignments in the dates were a coincidence. However, my essay dealt with the coincidence option in five different places, sequentially eliminating it as an unlikely scenario given the mounting weight of evidence for intention. The editors even addressed coincidence vs. intention in their preamble to my piece, indicating that this was something I must address (and I did). I pointed this all out to Aveni in an email of December 2013.

He didn't respond to my clarifying response, which was straightforward and civil. Through 2014 I tried several times to inform Aveni about ongoing publications (e.g., my Gelfer chapter and my peer-reviewed piece in *Zeitschrift für Anomalistik*, as well as my review of his new article, with the Brickers, on the

---

[19] For example at Colgate University (early 2012), linked on his website: http://anthonyfaveni.com/. Also his Penn Museum talk of 12-2012 (Youtube), where he claims my work is about the world's end.

Xultun tables).[20] But he was incommunicado. I also sent him my correction to his incorrect ballcourt alignment statement, in his 2009 book, but no response.

It's unacceptable when authoritative elitists refuse to acknowledge their errors and, in fact, just keep propagating their errors, unchecked, allowing their colleagues to continue citing them for assertions that are demonstrably false. I saw this happening in Krupp's 2014 article and 2015 preface, as well as in the Whitesides-Hoopes response to my critique of their paper (2014), where Aveni's book was cited as a reliable debunking of my work. One begins to suspect that a bully mentality underlies Aveni's otherwise cooperative-seeming persona. In fact, during an academic conference one of his well-known colleagues stated to me matter-of factly, without any leading provocation on my part, that Aveni is a "bully and a bullshitter."

By late 2014 I decided that I should ask his publisher to register an errata to the errors. Thus began one of the primary components of my Experiment, in early January of 2015. It was a new year, and I renewed hope that scholars and their allegedly reputable academic publishers might actually apply their own stated policies and abide by basic standards of science. So, the timeline in 2015:

1-5-2015. Query to Darrin Pratt
1-8-2015. Sent errors to Pratt / University Press of Colorado
Feb 2015. Their thumbs-down decision, via emails w/ Pratt
May 2015. Query re: Milbrath & Dowd book, sent to Pratt
May 2015. His response; also mentions that Aveni's "revised" eBook edition was
    released; additional exchanges follow
May-June 2015. Queries to AAUP
July 9, 2015. Complaint filed with AAUP
July 16. No response, query. Susan Patton acknowledges receipt
August 20, call Patton, she's on vacation
August 26, called Kyla Madden, left message. No response
August 30, called Kyla Madden, left message. No response
September 5, ditto, then response: material was sent to new chair of committee
September 10. Response is said to be forthcoming
September 29. "No basis" decision given by Peter Berkery
October 3. Email: Aveni admits his ballcourt alignment was a "mistake"
October 27-30. Exchanges with Peter Berkery. Checkmate and ... no response.

The University Press of Colorado is located in Boulder, Colorado, a place where I lived in the 1980s and 1990s. In fact, in 1997 you might say I worked for them — as a freelance reviewer for their *Colorado Libraries* magazine. It was an unpaid position, but I got to keep the books. I reviewed, among other academic books on Mesoamerican topics, Lindsay Jones's book on Chichen Itza, *Twin City Tales*. They have produced a fine array of publications for many years.

Anthony Aveni has been involved in many projects with the UP of Colorado. His offer to write a 2012 book, slicing out some market share of a pop culture craze,

---

[20] All three of these are at http://www.thecenterfor2012studies.com.

must have been enthusiastically welcomed. As it came to pass, Aveni's 2009 book *2012: The End of Time*, was the only book dedicated to 2012, written by a professional Maya scholar, that was published with a university press. This means it should have been subjected to peer-review, careful editing, and fact-checking. Its publisher has an obligation to ensure a reasonable standard of fact checking and error correction. All academic publishers acknowledge this, have a board of directors that oversee this, and often, as in the case of the University Press of Colorado, get their stamp of approval from the AAUP.

For years I had known that Aveni's book was filled with snide innuendos and errors, especially in regard to his treatment of my work. In 2014 I waded through the toxic mess again and selected seven unambiguous errors. These were not mere typos, although they covered a range of error types. I even included a fairly trivial one as a test, to see if the UP of Colorado would even acknowledge that one. Yes, I suspected that it was quite possible that they would deny everything, as with so many other efforts I'd undertaken to speak clearly to scholars about their misconceptions of my work and 2012.

I thought through my strategy and would hold to a standard of clear and professional communication. I had never communicated before with the press's Director, Darrin Pratt. My initial query was sent January 5, 2015:

Dear Mr Pratt,
I hope you can help me understand the policy of the University Press of Colorado, regarding errata — that is, factual corrections identified in your peer-reviewed books? Specifically, not just typos but mis-statements and factual corrections? What is your process for rectifying such errors in the published record? Do you typically deal with these in a second printing, or a later edition? I will greatly appreciate your assistance, and Happy New Year! Best wishes,

John M Jenkins

The next day he responded and explained that they would usually issue an errata sheet to be inserted into remaining stock, plus add a correction to future editions. I confirmed that I'd identified errors in one of their books, and emphasized that "these particular errors have impinged upon the reception of my own work, and at least one of them has been cited by other scholars as an erroneous precedent for dismissing my work. There's been a consequent chain of difficulties." I stated that they were in Aveni's 2009 book.

We agreed that an errata sheet would be an ineffective solution, as the book was virtually out of print. Pratt assured me we could "figure something out" and asked for my description of the errors. The next day I sent him my 9-page file, which can be considered the official "complaint" I filed. I don't like this term, as it doesn't capture the intentions of the process. It would be better to call it a "request for error assessment and correction." The file listed six errors on the first page, with the page numbers in Aveni's book where they occurred. A seventh error was detailed on page 8. After the error list, I explained the errors in more detail and provided corrective evidence, using quotations from my work and links. My cover letter, sent to Pratt along with my "complaint" is as follows:

Darrin,
I certainly hope so [that we can "figure something out"]. I'm attaching a word document. The errors are conveniently listed and cited on the first page. In my list I focused on explicit factual errors, and they could be re-phrased as an errata. In addition, it would be great if you or an assistant could read the extended narrative, which explains and supports the corrections and my other points. The narrative story drives home the problems better than the simple list.

This would be less of an issue if four scholars (in one trade book and two peer-review journal articles) did not cite Aveni's flawed assessments as a means of dismissing my work. I've added these sources at the end of the file; I think we could add another article written by John Hoopes (IAU Vol. 7, no. 278, 2011).

My unusual journey as an independent Maya scholar has brought me to the point of publishing a "review-essay" in a peer-review journal last year, with another being shopped around. The one publication was achieved with some difficulty. To have false statements about me and my work published in Aveni's book under the approval of an AAUP university press harms my publishing and teaching opportunities. Apart from this, simply on principle these errors should be corrected.

Thank for your consideration of these items and for wanting there to be a solution. Best wishes,

John Major Jenkins

I emphasized to Pratt how Aveni's unchecked errors continued to be cited, by other scholars, as viable critiques. This was an ongoing problem, damaging to the reception and continuation of my work. I made it easy for Pratt and his advisors to assess the veracity of the errors, which were tightly listed in less than a page:

1. Incorrect assessment of Grofe's methodology of approach, with a resulting thumbs-down assessment of Grofe's findings. Page 105.

2. Incorrect assessment of the direction of the rate of precessional change, with resulting incorrect assertion as to the inaccuracy of Grofe's findings. Page 96.

Note: Grofe's work provides support for the precessional premise of my 2012 astronomy work. The two errors above had deleterious repercussions in that they were cited and adopted by at least four other scholars, in order to show that my work was not viable or had been "debunked."

3. Aveni's supporting citation to a section of my 1998 book that does not, in fact, provide support for his demeaning assertion about my attitude toward Maya scholars and their critiques. Pages 23-24.

4. The use of religious bigotry as a critique: The false assertion that I belong to the religion called Gnosticism, packaged with his bigoted application of his belief, supposedly demonstrating that a Gnostic religious persuasion invalidates ones scholarly work. Pages 15-16, 18, 23, 158-159.

5. Factually incorrect statement regarding the Izapa ballcourt's horizon alignment, 48° in error. Pages 54-55.

6. Incorrect reporting of the McKenna brothers' book as a "booklet" published in "1971". Page 16.

As mentioned, a seventh error required more detailed explication, and is introduced on page 8. I'll start here, but be forewarned: this one involves complicated astronomy. Nevertheless, it's fairly easy to summarize the issue, and the problem with Aveni's assessment should be clear even to a non-astronomer. (I'll emphasize the issue in bolded font below):

There is one final comment in Aveni's book that I'd like to highlight for its misleading quality. I guess this could be called **error number 7**. This error is on par with Aveni's mistakes regarding Grofe's work. In discussing how the Maya might have been tracking precession, he writes:

Or you could track "one day of precession" by noting the slow shifting of the stars in the zenith, or the shifting dates of solar relative to stellar zenith passages, as some investigators have suggested[10] (104).

The end-note 10 refers to Appendix 3 and other pages in my 1998 book. The reference is to my identification of the sun and the Pleiades conjoining in the zenith as a central element in the New Fire Ceremony, which I discovered was also embedded into the architecture and orientation of the Pyramid of Kukulcan at Chichen Itza. This was unprecedented work, argued over several chapters in my book, though Aveni characterizes it as being "suggested" by "some investigators" (plural).

The problem here is how he dismisses it in the next sentence: "Once again, however, the problem is that stars shift at a variable rate" (104). **This is true for stars that are not on or very close to the ecliptic. My model utilizes the sun (which is by definition on the ecliptic) and the Pleiades star cluster, which is very close to the ecliptic.** Unlike stars far from the ecliptic, both of these shift at a constant rate with precession. The model involves the fact that the sun will be at the nadir when the Pleiades pass through the zenith *at midnight* (in November, which defines the New Fire Ceremony and the end of a 52-Haab Calendar Round). Consequently, I noted that this means the sun and the Pleiades will be in

45

conjunction exactly six months later, in May. The timing of this conjunction shifts with precession and slowly approaches the date of the solar zenith-passage in late May. The specific date of the solar zenith-passage is a function of latitude, but at Chichen Itza it targets a precessional alignment of "the sun and the Pleiades in the zenith" that is occurring in the 21st century AD — thus coordinating, via a totally different method of precessional calculation, with the Long Count's era-2012 galactic alignment.

This is heady stuff, but Aveni himself and professional peer-reviewers qualified to assess Aveni's book should have no problem understanding these astronomical dynamics. I mean, *I do*, and I'm suppose to be the uneducated Gnostic Y12er charlatan. If they fail to grasp how the precessional shifting of tropical anchors of the solar year along the ecliptic, in relation to background features such as the Pleiades, is not to be confused with variable shifts of rise times and azimuths at the horizon (as Aveni asserts), then they should take some astronomy classes. The problem here is Aveni's biased assumption of what should be measured, as MacLeod, Grofe, and I discovered in 2008 (see pp. 39-40 above).   Aveni misunderstands and rejects my reconstruction of a Maya method of tracking precession because it doesn't fit into his Western assumption of how it must be done. As I stated in my complaint: "Aveni's rationale for dismissing the relevance of this work is flawed because the solar and stellar features are not subject to the "variable rate" he states is "the problem" (that would only effect stars not on, or very close to, the ecliptic)."

The other errors are more easy to assess, and I think it's important that I summarize each one here. (The verbatim and unedited primary document is provided in Appendix 3; see p. 158 for link.)

First, although Aveni himself (with Hartung) measured the Izapa ballcourt's orientation, and published it in 2000, *he stated it 48 degrees in error in his 2009 book*. He got the orientations backwards, thus confusing the reader as to the relevance of an important facet of my reconstruction of Izapa's archaeoastronomy: the *December solstice sunrise* direction, pointed to by the ballcourt and the front face of its throne. He also managed to wiggle around acknowledging that I independently discovered this and was the first to publish it (in my 1998 book). This may seem trivial, but any reader, as well as his undiscerning colleagues, will read his words and get the impression that Aveni has corrected me and that my ballcourt interpretations are wrong. The problem is rooted in his *factually incorrect statement of the ballcourt's orientation*, compounded by his hostile attitude, his loose lingo, and his incomplete assessment of my work. He concludes by citing to Julia Guernsey's 2007 book on Izapa as the preferable go-to source, which in fact *doesn't even discuss the Izapa ballcourt* and its orientations.

Another error I pointed out involves a deceptive citation practice that undergrad students often commit, knowingly or just as a result of sloppy scholarship. More usually, what slips through and gets revealed is a prejudicial bias in the mind of the scholarly critic. They state a negative reading of a person's ideas, work, or attitude — which is nothing more than an imagined prejudice in their own minds — and then cite the person's work as proof of the critical assessment. Very few readers or fact-checkers will ever go to the cited source to confirm whether or not what is stated there affirms the critic's assertion. Most undiscerning readers who respect the authority and intelligence of a scholarly critic will accept the veracity of the critique, without even trying to confirm it. A problem occurs, however, when the cited source does not, in fact, contain anything that supports the critic's negative judgment. This is one of Aveni's errors, with a strikingly ironic twist. I'll just provide, below, a verbatim excerpt of the explanatory narrative that I included in my "Aveni error file," sent to Pratt on January 8:

The **third error** in my list involves Aveni's citation (pgs. 23-24) to a section of my 1998 book that *does not contain support for his assertions*; in fact, *it contradicts* his assertions. Aveni asserted a demeaning characterization of my attitude and reaction to scholars, which is not supported by the source he cites. Observe: On pages 23-24, Aveni stated that "Jenkins' ideas have not been well received among mainstream Maya scholars, who place little stock in subjective analogies and knowledge acquired through revelation" (implying, falsely, that this is my *modus operandi*). Aveni continues: "Meanwhile, freelancer Jenkins responds by disparaging the academic community of Mayanists who, he says, have shut him out and ignored him.[18]"

As support for his statement, the superscripted end-note 18 refers to Appendix 5 in my 1998 book *Maya Cosmogenesis 2012* (see the note on p. 168 of his book). If you would like to read my entire Appendix 5, I will send it. My quotes below make my point. Again, Aveni's assertion that I was "disparaging the academic community of Mayanists" and that I say I've been "shut out and ignored" is not supported. The issue here is that, in that entire appendix, *there is no statement or even an insinuation by me that I am complaining about being "ignored" or "shut out."* My appendix responds in a clear and cordial manner to Linda Schele's critique of 2012 ideas, including a few that are central to my work. My response is titled "Response to Counterarguments".

To heighten the irony, if we actually read my Appendix 5, which Aveni cites as support for his misleading comments, we find me supportively citing Aveni himself in my penultimate concluding paragraph (bolding added for emphasis):

If we do not allow these ancient skywatchers to have been sophisticated enough to notice precession, we relegate the alignment of 2012 A.D. to the unexamined bin of "coincidence." To conclude that this is coincidence pushes our thoughts beyond credible bounds of reason. The alternative, as resistant as many will be, is that the creators of the Long Count calendar calculated the rate of precession over 2,000 years ago. **Few Maya scholars are as qualified to comment on this point as archaeoastronomer Anthony Aveni, who wrote, "Ancient astronomers easily could detect the long-term precessional motion . . . Through myth and legend the earliest skywatchers transmitted their consciousness of the passage of the vernal equinox along the zodiac from constellation to constellation" (1980:103).** In the interest of clarity, I will mention that it would be more accurate to say that the alignment occurs in *the era* of A.D. 2012; because precession is such a slow phenomenon, fifty years on either side of 2012 might be appropriate. (Jenkins 1998:361).

To give a sense for the tone of my comments, the very first paragraph of my appendix reads:

This appendix grew out of a need to respond to "Comments on the Creation Date," posted on the Mesoamerican Archaeology Homepage website by Linda Schele, April 1996. The so-called "end-date" of the thirteen-baktun cycle of the Maya Long Count in A.D. 2012 has been the subject of internet discussions, and in early 1996 Linda Schele responded to a question regarding the Long Count. She addressed the importance the Maya applied to the date in A.D. 2012, and reiterated her viewpoint as found in *A Forest of Kings* (1990): "The Maya, however, did not conceive this to be the end of creation, as many have suggested" (82). This basically sums up her position on the meaning of the 13-baktun cycle end-date in 2012. The statement is essentially correct, because the Maya believed that time is cyclic, but continuing ambiguities demand that we clarify our terms and ask some more pointed questions regarding this date. I have always made an effort to refer to this date more specifically as "the end-date of the thirteen-baktun cycle of the Maya Long Count calendar," and I do not concur with the Neo-Atlantean pole-shift cataclysmologists on the idea that the world will literally end in 2012. With this distinction in mind, I admit that I still occasionally write, as a shorthand note, "end-date" or "end-date in 2012." This does not mean that I believe the Maya calendar or the world will end in A.D. 2012. However, as I will show, it is clear that the 2012 date was singularly important for the people who created the Long Count calendar (Jenkins 1998:357).

In reading these excerpts, does one get the same sense about my position regarding scholarly critics that Aveni painted, and cited to this appendix in my book? I really need to underscore this strongly: there is NOTHING in that cited source to support Aveni's misleading and disparaging characterization of me. AND, in fact, the appendix shows me engaging the critical counterarguments in a productive, informed, reasonable, and fact-based manner. — end of excerpt

So, what more can be said? This was in the material I sent to Pratt and, later, the AAUP, for their assessment. It's tragic and comic that the appendix Aveni cited shows me appreciatively quoting Aveni himself to logically support my idea about precession. My critique of Schele did not evince any angst or perception of being "shut out" or "ignored." Aveni employs a misleading citation practice that undergrad students learn, early on, is the sign of poor scholarship — although it can be intentionally used to give the illusion that your assessments of a person's work are supported by the cited reference. If this was, indeed, consciously done, then one can only conclude that an unethical maliciousness is at play, designed to mitigate.

Another one of Aveni's assertions about my background and work is both factually wrong and ethically wrong. Here we have the ol' *ad hominem* style of critique — which is directed "to the person." In other words, the critic attacks something personal in order to lower the standing of the person in the eyes of the reader. This is a cowardly work-around to dealing honestly with the person's ideas and arguments. If the person coming under scrutiny does have some controversial factor in their personal history, it's easy low-hanging fruit. Aveni does this well, and he does something here that is so heinous, so objectionable, that none of his reviewers, let alone his publisher, had the guts or discerning intelligence to point out. If truth and accuracy are important to scholarship, and I believe they should be, then it must be broached. Rather than re-narrate this error, I will simply quote from my complaint:

> A **fourth error** involves a false assertion as to my religious persuasion. In addition, we find what can only be described as bigotry applied as a critique. The evidence for this occurs over several pages of a constructed narrative, with several specific identifying statements in various places throughout the book. Early in his book Aveni constructs a framework in which various authors are framed as belonging to a belief system, and an approach to 2012 and the Maya, that he characterizes as "Gnostic." On pages 15-16, Aveni cites a scholar of Gnosticism (R. Grant) for a definition of Gnosticism, and writes:

>> One scholar has characterized Gnosticism as a mixture of eastern religions couched in the language of Greek philosophy and originating in "an atmosphere of intense otherworldliness and imaginative myth making." **These words fit today's 2012 wisdom seekers like a glove.** [emphasis added]

> That "today's 2012 wisdom seekers" include authors of 2012 books, such as myself, is suggested here *and then confirmed* in his following paragraphs in which he offers a definition of Gnosticism in a modern context followed by a sequential discussion if its purveyors. Speaking of the writings of the ridiculed "Y12ers":

>> Often laced with scientific language, this new brand of Gnosticism is built around the basic idea that all existences originate in a higher power that manifests itself by successive emotions that take the form of turning points, or turnovers, of eons.

> (Successive emotions?) His immediate sequential treatment of 2012 writers then begins with the preamble "let us take a look at some of these professed, latter-day wisdom seekers", showing how Aveni believes that the following authors belong to the category of "Gnostic" wisdom seekers. The litany includes, in sequence, Geoff Stray, Terence McKenna, myself (John Major Jenkins), Carl Calleman, Daniel Pinchbeck, Lawrence Joseph, and Whitley Streiber. He tacks on, for good measure, a few Maya "elders" and "seers".   —end excerpt

I go into further detail in my narrative, as Aveni reinforces this bigotry at other points in his book, alluding to me as a "Gnostic New Age prophet." But the situation is clear, and I summarized:

> The material I sketched above shows Aveni's various references to me as a "Gnostic", with false and forced arguments that associate me with dubious writers and concepts. His belief is that such a religious persuasion is highly suspect and conflicts with doing sound scholarship. First of all, I do not belong to the church of Gnosticism. I don't know how else to say this; I am simply not a member of that church, in the same way that I am not Jewish, Muslim, or Hindu. Thus his **fourth factual error** is his assertion that my religion is Gnosticism and I am a Gnostic. But what if I was? Since when does ones religious affiliation serve as a lynchpin for denouncing ones scholarly work? It is unclear how Aveni's bigotry has gone unnoticed and was given a pass. None of his reviewers have noted it; none of his fact-checkers or peer-reviewers caught it. Rather, it seems to have been allowed as a viable applied critique to myself and other 2012 writers. I don't think that bigotry is an acceptable method of academic critique. This should be considered another error worth addressing and, somehow, "correcting", but it is more egregious than a mere factual error — it is an ethical error and a violation of scholarly principles.

It may be that Aveni's flawed, factually incorrect, and ethically distasteful attitude, which I exposed and pointed to (like the kid saying that the emperor is obviously wearing no clothes) is so egregious, so vile, that it *must be* denied. It cannot be admitted to in any way that saves face, and the choice becomes clear: deny it and try to wiggle around the unethical violation of scientific principles. Or admit to it and *allow science to work*, to correct the balance so that facts and truth can be honestly acknowledged. Aveni and his publisher clearly took the former path.

I included a fairly trivial two-part error, as a test to determine if they were going to apply any kind of discernment to my complaint. On page 16 of his book Aveni described the McKenna brothers' *Invisible Landscape* publication as a "booklet" published in "1971." I have the first edition, published in 1975 by St Martin's Press. It is a full-scale hardback with dust-jacket. I offered this as an error that could be easily checked by practically anyone. In fact, I put the question to my friend's 13-year-old son, a good kid into sports and video games, and within two minutes online he provided the correct publication information. Would Aveni and his AAUP-approved academic publisher be able to eat crow, or even take just a little nibble, enough to acknowledge this simple test error? No, repeatedly, no, and no again.

After almost seven weeks of assessing and processing, I received a 167-word decision from Darrin Pratt:

Dear John,   February 18, 2015
After discussing this with valued advisors, I do not believe that issuing an errata sheet for Anthony Aveni's *The End of Time* is warranted. What you describe in your Word document detailing the purported factual errors may or may not, in

fact, be errors. Certainly the author does not agree that he is in error on these points, other than one point [Aveni's misreading of precession] I believe he has already confirmed with and that he is in the process of correcting.

Scholars disagree about minor and major issues all the time, and the usual scholarly way to handle this is to write your own article refuting the claims with which you disagree. In my opinion, and those of my advisors, that would be the correct process for the errors you identify here, rather than issuing an errata.

I know this is not the answer you wanted to hear, and I am sorry for that. Hopefully you will be able to find a venue to communicate your disagreements with Dr. Aveni. Best, Darrin

My response:

Darrin,
There are parts of your email that are contradictory and ambiguous, and I really need some clarification. We already agreed that an errata sheet would not be a meaningful solution. I might next suggest, noting your assurance that "we can figure something out" (email of Jan. 7) that you simply issue a statement, to me, that could serve as an official correction, but you state that the "purported" errors I identified "may, or may not, in fact be errors." So, you are saying that after six weeks you and your valued advisors were unable to determine with any certainty whether or not any of the seven "purported" errors are actually errors (apart from the one I noted that Aveni already acknowledged)? Is this correct?

Were your "valued advisors" legal consul or knowledgeable academic reviewers of the points?

It appears that you sent my list to Aveni for comment. Is this correct, and did he receive my entire file for review? And, just to be absolutely clear, you then state that he "certainly ... does not agree that he is in error on these points" (except for the one previously noted). This is information he conveyed to you after you sent him my entire list, with my fuller explication? Being clear on these points will be helpful as I take your advice and seek correcting the record through other channels. Best wishes,
　　　John Major Jenkins

Pratt replied briefly that, again, "My advisors do not agree that these are errors", and confirmed that Aveni had received both the list of errors and my longer narrative. His definitive denial of the errors certainly flew in the face of my simple test error (the McKenna book description and pub date). Since I had him on the line, I wanted to hold the mirror up to what he was actually saying and doing, to see if honesty and rational processing might be possible, as he tried to frame the issue as my "interpretations" versus Aveni's "viewpoint":

Dear Darrin,
This is not an issue of one opinion against another, or a "point of view." It's a question of verifying facts. In my email of January 8, I carefully selected seven factual errors, out of many baseless anecdotal assertions and presumptuous

insinuations that Aveni made in his book, which could perhaps be debated from various viewpoints. But I intentionally selected and presented to you factual errors.

So, as an example, it should be very easy for anyone, including yourself, to verify one of Aveni's factual errors. Google "Terence McKenna Invisible Landscape" and you find the Wikipedia page for Terence McKenna (http://en.wikipedia.org/wiki/Terence_McKenna). It verifies that *The Invisible Landscape* was published in 1975, not 1971 as Aveni stated, and was a full-length hardback book, not a "booklet" as Aveni stated. So, this is a two-pronged error. But your advisers "do not agree that these are errors" (and neither does Aveni).

You claim you must defer to advisers on these matters. But as a publisher, verifying such publication details, as in the example above, is indeed within your area of expertise. You probably have a publication database, or website, you frequently reference on related questions.

That your advisers were not able or willing to perform two minutes of due diligence to verify this particular error, and have directly misinformed you on this being an error, should be a red flag for you, as the Director of the University Press of Colorado, a member of the AAUP and for 50 years the trusted publishing arm of the University of Colorado. Your academic press is clearly employing fact-checkers and peer-review advisers who are either not doing their jobs very well or are biased.

Do you think you could take 12 seconds to click on that link above, scroll down to the Bibliography heading, and read the publication year of the first edition of the McKenna brothers' book? Thank you. You must now be wondering if your advisers were mistaken in their assessment of *the other* factual errors I identified in Aveni's book, correct?

Pratt dodged and reiterated that I should seek a solution elsewhere, which is a way that corporations wiggle out of responsibility for dealing with their malfeasance. It's called outsourcing. It is, in a word, *irresponsible*.

In a follow-up email (see below) I pointed out my simple test error, provided a link to the publication data on the McKenna brothers' book, and asked Pratt to click on the link. He evaded and waffled, saying that he was merely the publisher and re-asserted that his advisors and the author (Aveni) did not agree that any of the errors were, in fact, errors.

He suggested I take my grievance and publish it elsewhere, as if this was an effective way for me to deal with my "disagreement". I noted that this tactic was, essentially, like the way corporations use "outsourcing" to irresponsibly evade dealing with the toxic consequences of their own flawed policies. Furthermore, the obligation to deal with their inept fact-checking was displaced, or outsourced, onto me, whose work was the subject and victim of the errors. This could be understood as a form of "blaming the victim," well known in violent rape cases, of making the victim responsible for dealing with the damaging effects of the violation. They get abused and then are made responsible for dealing with it.

In any case, we don't have to speculate in this direction, for the situation is simple enough to understand: a scholar and his publisher refused to acknowledge, let alone correct, simple fact-based errors. They thus violate their own errata policy, which of course any legitimate university press publisher should uphold. We vaguely assume they will and always do, but we find out for sure only when they are confronted with this kind of situation. This is where the rubber hits the road.

After the initial round of email exchanges, totaling 17 emails through February 2015, no further progress was made, as Pratt was apparently placed between a rock and a hard place, between the corrective facts I presented and the denial of those corrective facts adamantly maintained by Aveni and their anonymous, hooded, "valued advisors." In terms of the basic values of real scholarship and academic publishing, for me his decision was unacceptable, in all good conscience.

My next chess-game strategy became clear. I had placed my effort on hold until early May, when I noticed that the UP of Colorado was publishing an anthology on Mesoamerican astronomy, which was basically on homage to Aveni. It contained several articles that mentioned 2012 or explored it in detail, and a preface by Ed Krupp — the Director of Griffith Observatory in Los Angeles who had repeatedly blamed me for the 2012 mess (see the Krupp section). So, I renewed contact with Pratt and asked if I could receive a review copy of the book, reminding him that I'd reviewed their press's titles in the 1990s, for the *Colorado Libraries Magazine*. I was concerned that Krupp or one of the other contributors (such as Aveni) might continue their factually flawed narratives about me and my work. If so, it would be yet another indication of how Aveni's book, uncorrected, continued to influence academic dismissals of my work. Pratt declined sending me a copy, but said that he did a search for my name in the book and didn't find anything.

Well, the truth is a little more complicated, but I'll reserve my discussion of that anthology for Chapter 4. What was interesting in our brief reprisal of May, is that Pratt offered the news that they had just released a "corrected" edition of Aveni's 2009 book, at least in the eBook format. I was surprised, as I'd not considered this as a solution to the problem of Aveni's errors. However, it seemed odd because the errors had already been repeatedly denied. Could they have reversed their adamant position? I asked to see it, and Pratt sent me a PDF, which was "last saved" on April 30, so it was fresh off the press, so to speak. The revisions, I hoped, incorporated at least a few of the needed corrections I noted. Pratt even said that it included a few of my corrections. But when I received and examined it, I was perplexed. Aveni had only addressed the Grofe error (in a brief added end-note).

Pratt had misunderstood some in-house communication, and clarified by sending me a missive from his production department:

Please note from production: May 11, 2015
"The highlighted Aveni text is attached here. The only change made based on Aveni's review of Mr. Jenkins' commentary is in note #12 on p. 173; the remainder are corrections that Tony himself requested a few years back."

That "change" on page 173 was Aveni's added end-note which addressed his misunderstanding of Grofe's work (see note, p. 64). But, again, that correction was actually planned by Aveni long ago, after his exchange with Grofe in late 2009. It was a coincidence that I'd also pointed it out, but Aveni wasn't addressing it because I pointed it out. The "remainder" of the corrections didn't involve the errors I'd articulated, but a few other unrelated minor points that Aveni had noticed and saw fit to address. They were all much less impacting than my own corrections, but Aveni had denied that my corrections were errors that needed correcting. See how this works? The factually flawed statements about my religious persuasion and my work on Maya precessional astronomy need to be preserved because they serve a purpose — that of mitigation. I've often felt that if I stated "2 + 2 = 4" scholars like Aveni or Hoopes or Krupp would do their best to disagree or otherwise deconstruct my statement. That's the corrupt scholarship of ego politics, employing poisoned polemics, intellect divorced from conscience or ethics, and the sorcery of semantics.

My final email to Pratt in this second round, which reached an impasse as intractable as the first round, maintained a professional reserve while insisting on asking the hard questions:

Darrin,

So, to be clear, this is a revision for the eBook only? And this revised version is now released & available in the market, as of this month? I imagine there may be a slight uptick in sales due to the many mentions of Aveni's 2009 book in the recently released *Cosmology, Calendars, and Horizon-Based Astronomy*, which is essentially an homage dedicated to Aveni's work.

FYI, the comment Aveni added on p. 173 is the long-needed response to Grofe's correction, which I understand Aveni was already planning to make. I had indeed reiterated that problem in my list of 7 errors, but it appears that none of the other errors I enumerated were addressed by Aveni, including the Izapa ballcourt error and the incorrect publication date for the McKennas' book — which are basic factual errors. Do you have an explanation for this oversight?

Are you, as Aveni's academic publisher, simply acquiescing to the demands or dictates of your author, rather than applying an objective peer-review assessment of their statements, as is your mandate as an academic publisher and member of the AAUP? Isn't the scholar/author suppose to be subject to the peer-review process facilitated by the academic publisher, rather than the academic publisher being subject to (i.e., controlled by) the dictates of the author/scholar?

John Major Jenkins

Pratt just kept going around in a circle of denial, and ended the exchange. At this point, I knew that I'd have to go to Plan B. Okay, I'll address "my problem" elsewhere. Having discovered that the UP of Colorado received its imprimatur from the AAUP, I noted that the AAUP upholds a standard of academic accountability via their Membership Standards and Policies committee. A complaint could be filed against a member publisher if they were not abiding by the required standards of academic publishing.

I was cautious and decided to make some general queries to the committee chair as to their policies regarding such matters. I confirmed who the proper contacts were, and that they independently assess a complaint, and don't defer to the judgment of their member publisher, or their board of directors. After some of the usual patient waiting through communication delays, on July 9, 2015 I emailed my official complaint to Kyla Madden and Susan Patton at the AAUP. It consisted of my original error file sent to Pratt on January 8, and a file containing my first round emails with Pratt (January and February). In my cover letter I summarized my communication with Pratt in May, and the missed opportunity for registering the corrections in the revised edition of Aveni's book. The subject line read: "Request for your review/assessment of an academic standards incident with the University Press of Colorado." My cover letter (take a deep breath, it's a long one):

Association of American University Presses (AAUP)
Admissions & Standards Committee
Kyla Madden, McGill-Queen's, Chair (kyla.madden@mcgill.ca)
Susan Patton (spatton@aaupnet.org)
Allyson Carter, Arizona
Gabriel Dotto, Michigan State
Garrett Kiely, Chicago
Leila Salisbury, Mississippi
Eric Schwartz, Columbia

Re:
Request for review of an incident with the University Press of Colorado, a violation of scholarly publishing standards and the terms of AAUP's admissions and continuing membership policies, as stated in Membership Guidelines.

Dear Kyla Madden and Susan Patton,   July 9, 2015

I communicated with you both last month regarding your academic standards policies. I appreciated your prompt responses to my questions. (I emailed through my account behak72, with the username Jon Behak, but I signed my emails John Jenkins. I don't want this to cause confusion. I am author, teacher, and Maya scholar John Jenkins (John Major Jenkins) and will hereafter use my other gmail account, the2012story@gmail.com, to communicate with you.)

I hereby formally request that you review the behavior of the University Press of Colorado, AAUP member publisher. The incident involves communications earlier this year with the UP of Colorado director, Darrin Pratt. It unfolded over a dozen or so emails, and he and his staff refused to acknowledge demonstrable factual errors in a book they published. The process I documented indicates their use of inept and anonymous fact-checkers, deference to the judgment of the author who perpetrated the errors in the first place, and their evasive request (after initially being proactive and cooperative) that I solve the problem myself in some other way, outside of their professional responsibility to directly acknowledge and correct errors in their own scholarly publications.

To be as clear as possible, and to help you understand this two-part incident, here is the sequence of events. The first attempt to seek a solution was in January-February 2015; a second communication occurred in May 2015:

January-February 2015
1. I contact Darrin Pratt and ask about their errata policy. Yes, of course, they do have one — an errata sheet in unsold book stock.
2. The errors are in a 2009 book that is now virtually out of print (Anthony Aveni's *2012: The End of Time*), so the standard errata-sheet policy would be ineffective. Pratt says, conciliatorily, that "I'm sure we can figure something out" (January 7th email).
3. I send the seven identified errors, easily stated in less than a page. Eight additional pages provide more details and corrective evidence, in a narrative form. I attach this file for your assessment, which was sent to Pratt on January 8, 2015; I also attach the entire email exchange between Pratt and myself, January-February 2015.
4. Almost six weeks elapse while Pratt's "trusted advisors" assess my alleged errors. Finally, I am informed that none of the purported errors can actually be confirmed as, in fact, being errors. They say that neither the book's author or their "trusted advisors" believe these are errors. I ask if their advisors are lawyers or knowledgeable fact-checkers. No response.
5. I am very surprised. Pratt is apologetic, as if he knows this is neither right or fair. A policy similar to "outsourcing" is asked of me, in which I am told I should take my grievance and publish it elsewhere. I take this as sloughing off, onto me, their professional obligation to correct errors committed by their author, which somehow made it through their peer-review academic publishing process.

(NOTE: one of my main issues, for you to consider, is whether or not the UP of Colorado is following an objective editorial process, or has more or less just published what Aveni handed them. As you can imagine, the latter scenario is open to abuse by axe-grinding scholars with an agenda to mitigate selected people. In Aveni's case, I have on record a first-hand account that he told his colleagues (at SAA 2008) to not refer to me or my books by name. This may have been his sour-grapes reaction after favorable coverage of my work in the *New York Times*, July 1, 2007 Sunday Magazine. So, there is a background and context for this issue with the ethics and standards of the University Press of Colorado — are they merely the voice-box of this senior author's personal agenda, or do they uphold academic principles and scholarly standards that override his personal agenda?)

I noted in my attached "error file" that two of the errors (pertaining to the work of Michael Grofe) were already acknowledged by Aveni; I was told by Pratt that Aveni was already planning a correction of some kind. Four of the other errors, as I explained in my file, pertain to my own work on Maya astronomy and period-ending beliefs relating to the controversial "2012" topic. These errors include *a demeaning use of my perceived religious persuasion as a means of critiquing my scholarly work.* I identified this as religious bigotry. The other errors involve a deceptive citation practice employed by Aveni (citing to evidence for his claims, where there is none), and centrally important aspects of my work on Maya astronomy and archaeoastronomy. I explained to Pratt that Aveni's book has been, and continues to be, cited as an authoritative "debunking" of my work, and therefore continues to be damaging to the accurate depiction of my work (see, e.g., some examples cited in my file). This is why I've sought a correction, and I stated to Pratt that I'd accept a written acknowledgment of the errors, with which I could attempt to remedy the existing and ongoing proliferation of mistaken judgments of my work, which authoritatively cite Aveni, in the published record and in high-profile information outlets such as Wikipedia. My attempt at a mild form of solution — a simple written acknowledgment of the errors — was rejected in no uncertain terms. Despite the fact that Aveni's book is on the shelves at over 500 university, technical, and college libraries, and will thereby continue to misinform students and readers indefinitely, I was willing to accept a mere acknowledgement of the errors.

The unprofessional behavior of the UP of Colorado is unacceptable, and thus I request that your Admissions & Standards Committee address this question:

Do you believe that the UP of Colorado is, in this specific incident, upholding basic standards of fact-checking (in the initial assessment and editing of Aveni's manuscript), error acknowledgement (after publication), and transparent and honest correction?

If yes, I assume that you have independently assessed errors 3-7 in my attached file (we can ignore the Grofe errors) and are not simply relaying some assurance garnered from Darrin Pratt. Then please provide me with the name(s) of your reviewers and advisors who cannot determine that errors number 3-7 in my file are in fact errors. If any of these can be found to be, in fact, an error (such as the publication year of the McKennas' book), then explain to me why the need for any correction was rejected by the UP of Colorado, and I was told, in effect, to get lost?

An addendum to this January-February sequence of events occurred in May, when Darrin Pratt informed me that a revised, "corrected" edition of the eBook format of Aveni's book was just released. But, upon examination, the only correction was in regard to Grofe's work, which Aveni was already planning on making. None of the other errors, which were of course already denied, were corrected. So, a perfect opportunity — one I hadn't even expected — to correct the record was passed by.

You explained to me in my earlier questions (from my gmail behak72), regarding member policies, that the scholarly qualification of each publisher cannot be determined by their in-house board members, and assessments of complaints take place solely in your office. I am not pushing for a 2/3rds vote to oust the UP of Colorado from membership in the AAUP. I am, rather, requesting an oversight review of this situation and some kind of corrective action to their unprofessional behavior. They've violated the primary mandates, stated in your Membership Guidelines (http://www.aaupnet.org/aaup-members/becoming-a-member/guidelines-for-membership), that AAUP member publishers must be concerned with *education* and following *scholarly standards*. The factual errors in Aveni's book do not educate, they misinform. The ugliest error is an ethical one of a bigoted, *ad hominem* narrative that denigrates a career author, teacher, and researcher. The *scholarly standards* in this book, and in the fact-checking of this book, are lacking. I have provided specific, tangible examples.

Here is the situation as I see it: The University Press of Colorado has produced *and then defended* a book that does not educate, but rather misinforms. A functioning and unbiased process by which misinformation (errors) should be corrected in a peer-reviewed academic publication was subverted, in favor of a senior contributing scholar (Aveni, who boasts in his book of having produced many previous projects with the University Press of Colorado). This is demonstrably shown, and easily shown, in the one test error that I included. It's a fairly trivial correction to Aveni describing a book, incorrectly, as a "booklet" published in 1971 (it was published as a full-scale hardback in 1975). Again, I included this on purpose, as a test. Any competent fact-checker employed by the UP of Colorado, or Pratt himself, could confirm this error online in less than two minutes. Anyone on the AAUP staff could Google the correct information. But this error, and all the others (apart from the two Grofe errors already noted) could not be confirmed as being, in fact, errors (please read Pratt's statements in the email exchange of February 2015). This strikes me as a gross failure of academic publishing and errata correction standards. I tried to resolve this with clear and honest communication, but that must work both ways, and Pratt has asserted the matter is resolved and he has nothing more to say. It isn't resolved, by any measure with which a functioning academic publishing & correction process should operate. So now I turn it over to you. **I assume you should be concerned that one of your member publishers is refusing to uphold the standards that your Member Guidelines require.**

In summary, the following issues have become apparent during this process, in which I asked the Director of the University Press of Colorado to assess seven factual errors that I had carefully selected in Anthony Aveni's 2009 book *2012: The End of Time.*

- Substandard peer-review and fact checking process.
- Apparent deference to the authority of book's author, rather than applying an unbiased and objective assessment of the errors.
- Refusal to even acknowledge basic, unambiguous, factual errors.
- Neglect in correcting the errors when the opportunity arose in later producing the revised eBook.

I'm willing to discuss this matter by phone, or to participate in a conference call, or answer any questions you have. I'm simply wanting a basic acknowledgement that the errors are errors. I carefully selected unambiguous ones amidst a morass of loaded lingo, snide insinuations and distorted paraphrases. I'm not asking you to address those; I'm asking you to address factual errors, each one of which is presented in clear language in the attached file (Errors-Avenis-Booka.pdf). After almost six weeks Pratt responded to it on February 18th with a 147-word dismissal. Best wishes,

    John Major Jenkins

Featured essays, interviews, peer-reviewed papers and chapters, at: *The Center for 2012 Studies* (http://www.thecenterfor2012studies.com). Bio: http://johnmajorjenkins.com/?page_id=16.

Attached files:
1. Errors-Avenis-Booka.pdf (sent to Pratt as Word-doc on January 8, 2015)
2. Gmail-DarrinPratt-Jan-Feb2015Policy-regarding-errata-AVENIsBook.pdf (emails with Pratt between January 5, 2015 and February 25, 2015)

After sending this, I waited a week with no reply. I sent a quick query and then received a brief acknowledgement from Susan Patton, that they were looking into it. I now shifted into extreme patience mode. Kyla Madden, who serves as a chief editor at McGill-Queen's, was featured in a panel of speakers at a publishing conference that I found on Youtube. In the peer-review process, they send manuscripts to peer-reviewers and expect a considered response after six weeks. Well, my complaint filing did require some focused reading, but nothing like an academic book-length manuscript. But I waited six weeks, and then asked for an update in late August. What steps had been taken to address my complaint? No response. So I left three phone messages for Madden and one for Patton over a two-week period. Finally, on September 8, a brief reply from Susan Patton: "I'm sorry I missed your call earlier. Just a quick update on the process. The chair of Admissions and Standards Committee is currently reviewing your complaint with the University Press of Colorado. We should have her recommendation on any further action shortly."

Then, a more accurate reply from Madden: "Thank you for your email and recent phone messages regarding your complaint about an AAUP member. My term as chair of the Admissions and Standards committee finished at the end of June, but I understand that this matter is under review with the new chair (Leila Salisbury) and that she hopes to have information for you soon. Thank you very much for following up and for your patience in awaiting a response." So, a change of the committee guard had taken place.

Given that this process was already hampered by delays, I decided to immediately establish contact with the new chair of the committee:

Dear Leila Salisbury,
After some persistent effort I was finally given an update from Kyla Madden, regarding the complaint I filed two months ago. She said she is no longer the

chair of the AAUP Standards and Admissions committee, and that you have taken on this role. Rather than repeat all of the basic orienting information regarding the issue I have raised and requested that the AAUP assess (regarding the University Press of Colorado), I wanted to simply be in contact with you to confirm that you were forwarded the cover letter and the files that I sent to Kyla Madden and Susan Patton on July 9, 2015. Have you received these?

In addition, since two months have now elapsed I am now inquiring as to what steps have been taken to review the matter? And I also want to extend my open communication with you, in the desire for a proactive solution, and let you know that I am available by phone or email should any questions arise. Phone 970 686-xxxx; email: the2012story@gmail.com. So, in summary:

1. Have you received the files previously sent, including the cover letter?
2. What steps have been taken, now that two months have elapsed, to address my complaint?

Thank you. Sincerely,   John Major Jenkins

She responded promptly: "I did indeed receive copies of all the materials that Kyla and the AAUP central office had received (including your cover letter). I reviewed the situation and sent an assessment to AAUP's central office for final review at the very end of August. I expect you should be hearing from the central office shortly."

Since Kyla Madden was my initial contact in this matter but was now out of the loop, and she was also a senior editor at a university press, I decided to invite some cordial discussion on religious bigotry. Had she any experience identifying bigotry in academic manuscripts, and to what extent might it be allowed or qualified? I know this was highly charged, given my critique of Aveni's religious bigotry, but I invited her insights anyway. On September 16 I emailed her:

Dear Kyla,
Thank you for letting me know. I was unaware that the baton had been passed to others. I was hoping for more of a dialogue than a decision handed down, case closed, but it's nice to know that something is moving forward.

In any case, I notice that you are a chief editor at a university press, and I'm sure you've had experience with wrong, slippery, or objectionable statements made by your scholarly authors. Do you feel that bigotry is easy to spot, or are there gradations of insinuation? Here's my question: Given the objectionable nature of such a tactic of critique — especially leveled against a living author — wouldn't you, as an editor, flag for clarification even the slightest possibility that a bigoted critique of a living author's religious persuasion and scholarly work was being made?

I'm doing a survey of editors and directors at university presses, and already have gathered over a dozen responses to this question. I was hoping to include your own response in my metric. Thank you and best wishes,
    John M Jenkins

Reasonable but challenging questions. Not surprisingly, she did not respond. And that is where the matter stood as of mid-September, some eight months after the process began. It was hanging in the balance. I was actually, believe it or not, hopeful that the system would work and that this astonishing situation would resolve itself with honesty and fairness. I had addressed a scholar, his publisher, and their supervising agency, with a clearly stated and supported complaint, on a matter of factual errors committed, with the expectation that all involved would practice their own policies and professional academic standards. If my request was to be once again thwarted or denied, then a certain interpretation gets a big boost, which would otherwise seem to be a baseless conspiracy theory. And that is: that my presence in the discussion of reconstructing Maya astronomy must be mitigated. The connection between me and certain ideas in the evolving field of Maya astronomy must be denied. But why would this be such an important precedent to establish at this juncture? Again, the most likely interpretation can be supported by the actual events I'm documenting here: it's because the ideas I pioneered and first published, which specifically relate to 2012 but also to reconstructing ancient Maya astronomy generally, are ones that professional scholars are now realizing are essential to progress in their field. We will see this most strikingly in Chapter 4, called "Ultimate Cognitive Dissonance."

As of September the Aveni debacle seemed to be pointing to either a partial victory, or to a total and utterly scandalous denial of the facts, the final jenga block being removed prior to the Ivory Tower house of cards crashing to the ground. I have not pursued redress through the legal system, as I believe these institutions have their own checks-and-balances that should function well enough. If they are totally resistant to acknowledging these errors and are vindictive enough, they may throw some lawsuit charge at me, intended as a gag order. I doubt that would happen, but if it does then this whole story will receive widespread mainstream exposure, and this book will document the facts and the truth. My goal is simple, and honest. I'm only concerned, here, with the factual veracity of Aveni's published statements, which were purportedly fact-checked and approved by his academic publisher, the University Press of Colorado.

Update. On September 29, 2015 I received an email from an AAUP Program Assistant, Bailey Bretz, with the brief message that I should read the attached letter. The letter was a PDF with an official AAUP header and signature of Peter M. Berkery, Jr., Executive Director of the AAUP.

Dear Mr. Jenkins,　　　　29 September, 2015
I am writing in response to your 9 July communication to Kyla Madden and Susan Patton requesting the Association of American University Presses to review an incident arising between yourself and the University Press of Colorado.

Upon referral to the Association's Admissions & Standards Committee, it has been concluded that the incident provides no basis upon which to take disciplinary action against the AAUP member press.

Thank you for bringing this matter to our attention. Cordially,

[signature written here]
Peter M. Berkery, Jr.
Executive Director

cc: Leila Salisbury, Chair, AAUP Admissions & Standards Committee
Susan Patton, AAUP Member Services Manager

And so, after almost nine months of persistent effort it comes down to, once again (like Pratt's response in February), a curt dismissal (even more curt than the previous one), effectively ignoring that there were errors in Aveni's book, errors that continue to be cited as valid and factually accurate debunkings of my work and career. These factual errors are not to be acknowledged, let alone corrected, in accordance with stated and known academic publishing policies. The Ivory Tower has officially become a House of Cards, with one scholar (Anthony Aveni) and his academic publisher, as well as the agency that supervises and mandates that publisher, insisting that they have the right to ignore and violate scientific standards of fact-checking and correction. How does science get broken? Scholars and their publishers break it.

The published record hovers with two different statements by Aveni, regarding the orientation of the Izapa ballcourt. The most recent statement of a scholar is usually taken to represent their updated or corrected position. I decided to contact Aveni for clarification:

Dear Tony,   October 3, 2015
You report the azimuth alignment of the Group F ballcourt at Izapa in two different ways in two of your publications. In your 2000 publication with Hartung, you have it as roughly 114 degrees. In your 2009 book *2012: The End of Time*, your words identify a 66 degree azimuth. You state:

> "Indeed, we found it to align approximately 1 degree off the December solstice sunset / June solstice sunrise direction" (Aveni 2009: 54).

An axis corresponding to "December solstice sunset / June solstice sunrise," as you state it, equals roughly a 66° azimuth at the latitude of Izapa. In recent correspondence with your publisher, I understand that you affirmed that this statement is not in error, and in fact it was not corrected in the subsequently released "corrected" edition of the eBook (May 2015). Am I to understand you have corrected your earlier work, and your position is now congruent with the latest statement, from 2009? And your calculation published in 2000 was in error? I'd appreciate some clarification of this murky issue. Thank you,

John Major Jenkins
Maya Cosmogenesis 2012 (1998)
Galactic Alignment (2002)

The 2012 Story (2009)
Director of *The Center for 2012 Studies* (http://thecenterfor2012studies.com)

Aveni replied later that day, without greeting or signature: "It is 114 degrees 29 minutes, originally reported in Aveni and Hartung 2000 "Water, Mountain, Sky" IN CHALCHIHUITL IN QUETZALLI ED. E. Quinones Keber, p. 60. Sometimes people murk mistakes."

Here Aveni affirms his earlier position, which means that his later statement, in his 2009 book, is in error. This is as close as Aveni could come to acknowledging his mistake, which he refers to as a "mistake" in his email above, trying to be cute by using the word "murk." But throughout this whole process, Aveni, his publisher, and the AAUP consistently refused to acknowledge this mistake, let alone implement a correction of it. I responded with a direct question for Aveni:

Tony,   October 3, 2015
Okay, thank you. After communicating this mistake and several others to your academic publisher, Darrin Pratt at the University Press of Colorado, he replied to me on February 18:

> "After discussing this with valued advisors, I do not believe that issuing an errata sheet for Anthony Aveni's *The End of Time* is warranted. What you describe in your Word document detailing the purported factual errors may or may not, in fact, be errors. **Certainly the author does not agree that he is in error on these points**, other than one point I believe he has already confirmed with and that he is in the process of correcting [the Grofe error]."

And, later, the error wasn't addressed or corrected in the revised eBook (which was released in May). Why did you deny this was an error at that earlier time, and neglect correcting it when the opportunity arose for the revised eBook?
John

And … wait for it … no response. And what if I was to present Aveni's confirmation that his 2009 statement was in error to the AAUP? Well, let's find out, but let's bide our time and play this out strategically. I began by asking Leila Salisbury, the new chair of AAUP's Committee for Member Standards and Policy, if her committee had tried to verify, independently, the errors I pointed out:

Dear Leila Salisbury,   October 5, 2015
It appears you were cc'd on this communication to me from the AAUP office of Peter Berkery, regarding my complaint with the University Press of Colorado and AAUP. There seems to be a piece of intelligence missing in this communication. I wasn't asking for a disciplinary action to be taken. I was asking that your committee, according to one of your purposes, assess and confirm or dis-confirm the errors I indicated, and which the University Press of Colorado refused to acknowledge, let alone correct. I had already previously confirmed with Susan Patton that your committee would conduct an independent review and assessment

of the errors. What is the result of this assessment of the errors under question, which I enumerated in my complaint? A decision to discipline or not is a separate matter. You said to me in a previous email that you had assessed my complaint and had sent a recommendation on to the central office (apparently, to Peter Berkery). The piece of information that was not communicated to me by you or his office, which was really the only thing I was interested in, is whether or not your committee determined that any of the errors I pointed out, and provided corrective evidence for, could be verified as being, in fact, errors.

So, I have two questions: 1) Did your committee in fact conduct this assessment, independently of assurances or opinions offered to you by the author of the book (Anthony Aveni) and the University Press of Colorado (Darrin Pratt)? **2) Did you determine, in your investigation, that any of the errors I enumerated are in fact errors?** The answer to question #2 is primarily what I'm interested in, and it seems that your committee must have arrived at a position on this question.

As I expressed in my previous communication, I'm not interested in disciplinary action (e.g., a vote to oust the UP of Colorado from membership), but only in your professional, objective, checking of the errors I identified. This is the crux of the issue, since my "purported" errors were expressly denied as being errors by both Aveni and Pratt (except for the Grofe error previously acknowledged by Aveni). Could you please clarify this for me? Thank you,
John Major Jenkins

The critical thing is that they simply said no disciplinary action would be taken, rather than admitting the errors and providing me with the results of their investigation. This is typically how guilty parties wiggle around culpability. It's what lawyers advise guilty people, businesses, or corporations to do.

I was holding Aveni's communication like an ace up my sleeve. I felt that the AAUP was now in a checkmate situation, because there were two options, both of which meant they were not performing their function. First, they may have determined that one or more of my errors were, indeed, errors. In this case, it's a very large question as to why no "disciplinary action" would be taken. Second, they may assert, like Darrin Pratt before them, that it was not possible to determine that any of the errors were, in fact, errors. In this case, we have Aveni's email that confirms the Izapa ballcourt statement he made was not correct. It may be that, while no disciplinary action was taken, they did ask Pratt to make corrections or errata. Because of the AAUP's total lack of full communication with me, that would remain a chess move they might invoke, after the fact, should my continued efforts bring them to an impasse. What I'd like to know, right now, is whether or not they checked the errors — that was, after all, the purpose of the Committee for Member Standards and Policy. But Salisbury, in her response to me email, dodged providing a clear answer:

Dear Professor Jenkins,   October 6

I believe what you are requesting is outside of the scope of my committee, and I'm referring your query to Peter Berkery in the central office. With best regards, Leila

This was the classic move of referring to a higher authority — the central office. But Peter Berkery couldn't be expected to have detailed knowledge of the situation; he might not even have read my complaint. His role was to read Salisbury's report and "recommendations", and make an executive decision about any actions to be taken. Salisbury's committee had not been forthcoming with the steps they'd taken to address my complaint, which is what I inquired about in early September (after they'd had my complaint in hand for almost two months). I explained the contradiction in a response to Salisbury:

Leila,   October 6
I'm afraid this sounds like I'm getting the run-around. As I said in my recent email to you, I had already confirmed with Susan Patton that your committee is tasked with independently and objectively assessing the merit of my complaint — and this of necessity would involve confirming or dis-confirming the "purported" errors I pointed out. Do you not have any write-up or report on hand, from fact-checkers you employed for this purpose? If so, I would like to see it. If not, then you did not perform the objective evaluation which your committee is responsible for.

I'm asking a simple question: did you (your committee) confirm that any of the errors I pointed out were, in fact, errors? Would this not be the primary factor upon which any disciplinary decision would be made?

At the very least, I would like to see the "recommendation" you sent to the central office, upon which they based their decision. You alluded to this in a previous email to me in early September. This must have included your committee's evaluation of the merit of my "purported" errors. Otherwise, the central office could not have been adequately informed during their decision-making process. Thank you,
    John Major Jenkins

After three days and no response, I tossed a final effort in her direction:

Leila,   October 9, 2015
How can I get an answer to my query? I'm honestly and clearly asking for the results of your assessment of my complaint. The process of checking on complaints was made clear to me by Susan Patton. I understood that your committee would objectively and independently assess the merit of my complaint, which involves the factual errors I pointed out, and which the University Press of Colorado denied were verifiable errors. The reason I brought this to the attention of your committee at AAUP, is because I confirmed (earlier, with Susan Patton) that it was within the jurisdiction of your committee to

ascertain whether or not the errors were actually errors. Whether or not my complaint has any merit is predicated on that question. What did your investigation determine? The brief "decision" conveyed to me on September 29, from the AAUP central office, merely tells me that 'we are not going to do anything about this.' But it doesn't tell me the results of your investigation. I assume that during the several months during which I waited for a response that your committee performed an investigation, as that is the function of your committee. I'd appreciation some guidance on this matter. Sincerely,

John Major Jenkins

To this she briefly responded:

Dear John,   October 9, 2015

I understand that this may be frustrating to you, but I'm going to have to refer you to Peter Berkery at the AAUP's central office for any additional information about how the matter of your complaint was handled.

With best regards,
  Leila

Okay, so she claims that Peter Berkery will be informed on the matter of how my complaint was handled. I predict he will evade sharing any information regarding whether or not the errors I identified were checked, and whether or not it was determined that they were actually errors. Because, again, either response to these yes-no questions proves malfeasance on the part of AAUP. Let's see how this goes. First, let's get to know Peter Berkery. He is the Executive Director of the AAUP. His background and credentials are provided on his AAUP page:[21]

Peter Berkery has been Executive Director of AAUP since early 2013. Berkery comes to AAUP from Oxford University Press, where he served for the previous five years as Vice President and Publisher for the US Law Division. Prior to that he worked for Wolters Kluwer for 11 years in a series of positions, publishing works on securities licensing examination training, securities law, taxation, and financial planning. He began his publishing career at a division of Thomson Reuters.

Berkery has extensive experience in government affairs and association management. He has been Director of Government Affairs for the National Society of Accountants and Government Relations Counsel for the National Paint and Coatings Association, and has served as Assistant Executive Director and Staff Counsel for a division of the American Trucking Associations. He has served on the Board of Directors of the Accreditation Counsel for Accountancy and Taxation, and as its President.

---

[21] http://www.aaupnet.org/about-aaup/aaup-staff/aaup-executive-director.

Berkery has a BA in Classical Studies from Boston College, and both an MA and a JD from The American University, as well as a Master of Laws in Taxation from George Washington University. He has been admitted to practice in Maryland, the District of Columbia, Hawaii, and the United States Tax Court. He is a certified financial planner.

His email and phone number are freely posted on the AAUP website: pberkery@aaupnet.org; 212-989-1010 x29. It feels a bit like we're going down the rabbit hole here. US Law, taxation, financial planning, VP of the US Law Division at Harvard University Press, business management — can I expect that he is even familiar with the details of my complaint? My best approach is to request from his office the recommendations from Salisbury's Committee, which she refused to share with me and upon which his decision was based. I'm trying to determine if they found that 1) none of the errors are errors, or 2) one or more of the errors are errors. If the former, Berkery's decision is understandable, but then we have Aveni's recent admission of his mistake which contradicts such a finding of the committee. If the latter, then Berkery's decision is less understandable, and some explanation should be offered. A third possibility is that it doesn't matter whether or not there are errors, they aren't going to do anything about it. In this case, the AAUP is functionally broken, its "standards and policies" have no meaning, and the AAUP is therefore in violation of the principles of science, principles which they claim to follow and uphold in their stated policy.

I took the next step and called Peter Berkery. It was not clear from the "decision" whether or not Salisbury's committee had indeed performed its assessment function. Did they determine that all of the errors were invalid? I would have to get that answer through the Executive Director of the AAUP, Peter Berkery. On October 27 I called his office but got the voice message. In it, he said it's better to contact him by email. So I didn't leave a message, instead sending the following email:

Dear Peter Berkery,   10-27-2015

I was directed to you by Leila Salisbury, chair of the Admissions & Standards Committee at AAUP, for an answer to a simple question that she was unable or unwilling to provide.

I filed a complaint with the AAUP back in July, regarding the refusal of the University Press of Colorado to acknowledge, let alone correct, a variety of errors in one of their books (*2012: The End of Time*, by Anthony Aveni, 2009). These errors had, and continue to have, deleterious repercussions in how my work has been viewed by other scholars in the field. In turning the matter over to your AAUP member Standards Committee, I was asking for an independent and objective assessment of the validity of the errors I had enumerated.

A "decision" was sent to me from your office on September 29, stating that "the incident provides no basis upon which to take disciplinary action against" the University Press of Colorado. I had explicitly stated in my cover letter to the committee that I wasn't interested in any kind of punitive action, such as a majority vote to oust the member press, but merely in knowing whether or not they could confirm, objectively and independently, that any of the errors I pointed out were actually errors. Before I filed my complaint I had previously confirmed, in a communication with Susan Patton, that it was the function of the member standards committee to perform this assessment, outside of any influence from the member press under consideration, or their board members.

In stating that there was "no basis" for any action, your decision certainly gives the impression that the assessment was indeed performed and none of the errors could be verified as being errors. But I didn't want to presume, so I sought a simple clarification with Leila Salisbury, asking the simple two-part question:

Did the committee assess the various errors I articulated in my complaint, and did they find any of them to be valid?

It's basically a simple yes/no question. I'm not sure why the chair of the committee that supposedly read and analyzed the merit of my complaint could not answer this question, but instead directed me to speak with you. I would therefore like to make an appointment to speak with you by phone. I can be available at any time during your office hours; five to ten minutes will suffice. Please suggest a good time that I can call and reach you. Best wishes,

John Major Jenkins

On the evening of October 28, Peter Berkery responded with the following:

Dear Mr. Jenkins,

Thank you for your email. I apologize for the confusion regarding the nature of your query to the association regarding the University Press of Colorado. It was never clear to anyone in the Central Office or the Admissions & Standards Committee that you were requesting the assessment you describe below. The association does not perform such assessments. Moreover, the deliberations of the AAUP Admissions & Standards Committee are confidential, and it would not be possible for me to disclose the nature of those deliberations to you. AAUP's sole interest in question's regarding a member's conduct is whether it calls into question a press' continued eligibility for AAUP membership – a matter exclusively internal to the association.

While I am happy to schedule a time for us to speak via phone, in fairness I must advise you that our conversation will be constrained by the restrictions I've just described.

Kind regards,

Peter Berkery
1-212-989-1010, x29 (publicly published)

I received and read it the next day and after some thought I responded:

Dear Peter Berkery,   October 29, 2015
I was quite clear in why I was requesting that the AAUP Admissions & Standards Committee assess my complaint, and in my emails of June I offered to further discuss and clarify any questions that the Committee might have.

I'm not sure if you are aware of the details of the complaint. One of your member publishers refused to acknowledge, let alone correct, factual errors in one of their publications — most of which have serious implications for the accurate reception of my work and background. It is a fundamental responsibility of legitimate academic publishers to recognize errors and offer corrections for the published record. Their refusal to do so would call into question their continued eligibility for AAUP membership. (Any "disciplinary action" to be taken is a separate question and not my concern; I was simply asking the AAUP Admissions & Standards Committee to perform one of its stated functions.)

The merit of my complaint is predicated upon the validity of the errors I indicated, and thus an assessment of those errors would of necessity need to take place. I had confirmed earlier with Susan Patton that the A&S Committee performs this assessment independently and objectively, and it's clear from the online mission statement that such oversight is one of the functions of that Committee. It is a check on the "standards" of admission and continued membership.

But you just informed me that the AAUP does not perform such assessments. You wrote: "The association does not perform such assessments." This is a contradiction. Furthermore, if the merit or validity of my complaint was not actually assessed, how, then, can you determine there is "no basis" for any action, as your letter of September 29 informed me? Upon what was that decision based, if not upon an assessment of the merit of my complaint (the validity of the errors I pointed out)?

Based on what you've told me, the following are therefore correct statements:

Your decision that there was "no basis" for acting (letter of September 29) occurred without actually assessing the merit of my complaint (the validity of the errors).

It is possible for one of your member publishers to violate principles of academic publishing without any oversight, assessment, or action taken on the part of the AAUP. They can maintain their membership in the AAUP without a concern for violating the foundational principles of academic publishing, because the "standards" Committee does not actually perform assessments of their behavior when complaints arise.

Thank you for clarifying what the AAUP does and doesn't do. Sincerely,

John Major Jenkins

On the morning of October 30, Berkery replied:

Dear Mr. Jenkins,

As much as I think it would be better all-around to let this matter lie, I feel obligated to correct an error of fact which has informed your conclusions below ("the following are therefore correct statements") ...

The AAUP does not perform the type of editorial assessment you requested of us for third parties. As our Admissions & Standards Committee (mis)understood your initial outreach to be a complaint against a member, the committee of course reviewed the facts you presented. However, as I explained, they did so in the context of determining whether or not they established grounds for disciplinary action against a member – their sole responsibility in assessing the facts you presented. Because member discipline is an exclusively internal matter, as I explained previously under no circumstances would that assessment be shared outside the organization. And, at the risk of being redundant, the context in which the committee assessed the facts is materially at variance from the context in which you requested an assessment (i.e., disciplinary action vs best practice). So, an assessment occurred, it just was not for the purpose you requested.

Needless to say, without elaboration the association therefore would reject your two "correct statements", the latter viscerally.

I hope this clarifies our actions on this matter. I apologize if my prior response contributed to a misunderstanding of our position, which I now trust is fully resolved. I wish you well in your future endeavors.

Regards,

Peter Berkery

I immediately composed and sent the following response:

Dear Mr Berkery,

So, I see now that the AAUP (standards committee) does indeed perform assessments of complaints, and did in fact assess the errors I presented. Thank you for the clarification.

Since it was found that there was "no basis" to establish "grounds" for any disciplinary action to be taken against the member press (the University Press of Colorado), it therefore follows that none of my alleged errors were confirmed as being errors. Correct? Otherwise, the alternative scenario is that no disciplinary action was taken despite one or more errors being verified. And if this is the case, then my second statement (in which member presses are free to violate academic publishing principles and behave unprofessionally without fear of discipline) would indeed be a correct statement. But since you viscerally reject the correctness of that statement, then I am left with the conclusion that NONE of my alleged errors were confirmed as, in fact, being errors. So, my revised statement of fact, based upon your clarification:

The AAUP Admission & Standards Committee did perform a review of my alleged errors, and found that none of them were valid.

Unless you inform me otherwise, I will understand that you agree with this statement, as it proceeds logically and directly from what you've communicated. Sincerely,

John Major Jenkins

He immediately replied:

Mr. Jenkins: as I have explained twice now, the assessment of the Admissions & Standards Committee is completely confidential. While I cannot agree with your statement, I therefore cannot comment further.

By any reasonable reading of this exchange, it is clear that the AAUP cannot provide what you seek. I do not wish to engage further in a correspondence that assumes my silence confers assent, so I will end this matter here.

With kind regards,

Peter Berkery

I immediately responded:

Dear Mr. Berkery,
I think the more accurate way of stating this is that "the AAUP **will not** provide what [I] seek." But I have my answers now. What I've sought is the answers to a simple two-part question: did you assess the merit of the errors I identified and did you find that any of them are valid? You have now answered the first part, and thus logically follows two possible answers to the second part ("yes" or "no"). Unless you can offer some other semantic gymnastics to wiggle around the facts, either answer leaves the AAUP in a difficult position.

If "Yes", then the correctness of my second statement in my previous email, despite your visceral disagreement with it, is indeed correct. If "No", then you essentially disagree with what I recently learned from the author, Anthony Aveni. In an email communication with me of October 3, I asked him to tell me which of his statements about the Izapa ballcourt alignment was correct — the one in his 2000 publication (w/ Hartung), or the statement in his 2009 book. As you'll recall, I had noted that his 2009 statement was incorrect by 48 degrees. A simple matter of acknowledging and correcting — but one which has taken over nine months to address, with denials and evasions all along the way, up until the present moment where it remains, in your hands, uncorrected. Aveni and the University Press of Colorado denied it (in February), but now Aveni has let slip an acknowledgment that his 2009 statement, in his UP of Colorado book, was a mistake.

So, having informed you that the original author admits to an error, I hope you will take steps to speak with or discipline the University Press of Colorado for poor fact-checking and subsequent evasions and denial of a correctable error in one of their books. Otherwise, as a legitimizing agency the AAUP is functionally broken. It's now proven that the UP of Colorado essentially violated scientific principles of reputable academic publishing (namely, fact-checking and honest error correction). The AAUP was subsequently complicit in not adequately assessing or requiring a correction of the situation. Thank you for wishing me well on my future endeavors, one of which you will be hearing about quite soon. Checkmate,

John Major Jenkins

This process apparently could continue going around in circles, *ad infinitum, ad nauseum.* I think that "disgusting" is a pretty accurate concluding assessment of these events. This is the preeminent iconic example in my Experiment. It's taken almost ten months of persistent and patient effort to bring it to some semblance of completion, with my complaint filed to both the University Press and then, that failing, to the AAUP (the professional association that supervises and validates them). One begins to understand that academic principles are a lower order of value to both personal ego and legal concerns — that a university press director will be advised, legally, to not admit any errors or wrong doing. Thus, scientific progress cannot happen because of the fear of legal action. But I never threatened legal action — I only requested academic honesty and accountability.

The problem seems rooted in the individual or individuals involved. At some point in their career, a certain scholar confers themselves with such unassailable grandeur of authority that their mere opinions and knee-jerk judgments should be received as gilded fact. Truth schmooth, and facts be damned. Their actions destroy the very foundations of their careers: the values of the scientific process. Like a smooth politician, and armed with a higher IQ as well as publishers who will defend them, they can slither and wordsmith their way out of any difficulty. They see themselves as being too smart to fail, and that is precisely what makes them too dumb to win. It was perhaps a bit brash of me to end my exchange with Peter Berkery with "Checkmate" — but that was indeed what his comments logically led to. I have left him to stew in his own evasive malfeasance. He maneuvered himself into a corner, and exposed the logical contradiction of the AAUP's handling of the situation. The payoff is that it's all documented, here, in my book, for all to see. The one peer-reviewed University Press book on 2012 written by a Maya scholar is an error-riddled fiasco defended by a functionally broken system.

◊ ◊ ◊ ◊ ◊

Ed Krupp: The Folly Conductor

As with Aveni, my communications with Dr. Edwin Krupp go back to 1996. It was at that time that I was inviting scholars to receive and critique my new work on 2012 astronomy, and was seeking an academic publisher for it. In my naiveté, I believed this was a possibility, but soon discovered that every university press I contacted reacted immediately in a negative way to a book about 2012. My proposal was cut short, phone calls ending cordially enough, but their blunt disinterest had nothing to do with my status as an independent, non-degreed autodidact, because the phone conversations never got that far. To the academic gestalt of the day, "2012" was simply not a legitimate topic for rational investigation.

Scholars like Krupp were infused with such suspicions. To his credit, he did apparently read and respond to my initial letter and articles (in 1996), and my follow-up letter as well (1997). But his comments seemed a bit flippant, like he just skimmed over my written words. His objections, too, were oddly irrelevant and evaded the actual evidence I cited. For example, he stated that there was no evidence that the ancient Maya considered "the star field of Sagittarius" to be a Creation place. Well, the constellation of Sagittarius is where the southern terminus of the Dark Rift in the Milky Way, the Crossroads of the Milky Way and the ecliptic, and the nuclear bulge of the Galactic Center are located — that was the point of my observations. And the Dark Rift and the Crossroads, at least, could definitely be shown to be central players in the Maya Creation Myth (citing Dennis Tedlock's notes to his translation of the *Popol Vuh*). That seemed worth talking about. But Krupp, like Vincent Malmström years later, dodged around that part of my argument, and the evidence I cited in support of it.[22]

Those initial back-and-forth letters at long intervals stretched into 1998, when I sent him my 1998 book for review in the *Griffith Observer* newsletter, which I was informed was a distinct possibility, though he was currently busy with summer programs. I waited until early 1999 and sent a query with some more detailed comments on the evidence that underlay my work,[23] but to this letter I received no response. That's the last I heard from Krupp until I reinitiated contact with him over sixteen years later, in May of 2015. But he certainly had not been idle in his "reviews" of my work, although a review never appeared in the *Griffith Observer*.

The next time Krupp came to my attention was in October of 2009. The *Sky & Telescope* magazine announced their next month's featured article — called "The Great Doomsday Scare" by Ed Krupp.[24] The brief description on their website was predictably loaded with innuendo and judgment, and I added a comment to their page, offering to write an article about serious research being done to reconstruct ancient Maya astronomy as it relates to 2012. No response. A few weeks later I read Krupp's *Sky & Telescope* article and was so shocked at his ridiculous

---

[22] See my 2002 IMS article and 2005 emails with Malmström (Appendix 1 online, p. 158).

[23] This letter formed the basis of my "Open Letter to Mayanists and Astronomers" which I posted to the Aztlan e-list group in June of 1999, and which then became an article published in the IMS *Explorer* in 2002.

[24] This was later posted to both the NASA website and the Griffith Observatory website.

73

assertions — essentially blaming my 1998 book for initiating the Great Doomsday Scare — that I wrote a lengthy corrective review-essay. I posted it to my new website, called Update2012.com, and sent links around to e-lists, friends, and Maya scholars. Eventually Krupp's article was posted on both the NASA website and the Griffith Observatory website, where it finally represented Krupp's promised "review" of my 1998 book, offered more than 16 years earlier. An "updated" version of the article was published in *iQ Magazine* in late 2012, which merely continued propagating existing errors as well as asserting a few more.

Returning to the events of November 2009, my book *The 2012 Story* had just been published and the next six weeks involved endless radio shows, newspaper and magazine interviews, and bookstore appearances. A lot of this, in the media, was driven by the doomsday-2012 movie that had just come out, directed by Roland Emmerich. To my surprise I was invited to speak to the media about 2012 at the Sony Pictures press conference. I walked down the red carpet and attended the premier in Hollywood with my beautiful wife, Ellen. It was certainly an exciting experience, and allowed me to point to my new book as the go-to place for the truth about 2012. In other words, I didn't kowtow to the Hollywood doomsday machine. I said in no uncertain terms, speaking to the media on the red carpet and during other interviews, that the Maya didn't in fact predict the end of the world in 2012.[25]

Little did I know that, on that very same day that my wife and I were dodging the doomsday media blitz on the red carpet and I was doing my best to get the truth across, Ed Krupp was nearby, speaking at The Beckman Center. His presentation was sponsored by the AIA and other scholarly societies, and was basically an elaborated version of his *Sky & Telescope* article. It contained a few gems that his article did not, which more deeply drove in his rhetorical imputation that I was to blame, more than anyone, for the "current flurry" and "character" of the 2012 mess (a.k.a., the doomsday scare). He affected this deception through another great tactic of the malicious inquisitor: selecting a truncated quote from my book and cleverly re-framing it to give it a pejorative slant. His manipulation of my actual written words was so clever that I can't believe it was accidental, and he repeated it again and again in his subsequent presentations, at other academic venues, for several years.[26] You, the reader, can judge for yourself.

But before I get too far ahead of myself, I want to emphasize that I was unaware of Krupp's presentation at The Beckman Center until I encountered it, in full, online in 2014. Presentations where "anti-2012" conferences had occurred were just recently being uploaded to Youtube and other websites connected to scientific institutions. I was astonished, in 2014, to see a tapestry of attacks on my character and my work, leveled against me by Krupp, Morrison, Hawkins, and other scholars at reputable scientific venues. Krupp gave virtually identical performances of his Beckman Center talk at other venues, such as his alma mater, Pomona State College (in 2010).

---

[25] The other invited participants in the Sony Pictures publicity campaign were Lawrence Joseph and Daniel Pinchbeck, along with Elizabeth Upton and History Channel producer Sarah Hollister. See www.youtube.com/watch?v=tV-tOkWmF1Y

[26] Pomona College, Beckman Center, Griffith-NASA event in December 2012 — all online.

I reviewed in detail what he did and said in his Beckman Center talk, and prepared a document for his consideration in which I laid out several questions for him to consider. But time passed, and by early 2015 I was busy playing multiple chess games with the Aveni and Morrison components of my Experiment. And then two threads came together. As previously mentioned, in May of 2015 I noticed that the University Press of Colorado was publishing a new anthology of academic writings on Maya astronomy. After Darin Pratt declined sending me a proper review copy, I checked out the book at the CSU library in Fort Collins. I got it as they were processing it before shelving, fresh off the press, and they even gave me the dust jacket (which they usually just toss). The dust jacket picture provides a hilarious confirmation of what I call the "Ultimate Cognitive Dissonance" of 2012 — see Chapter 4.

Maya scholars John B. Carlson and Clemency Coggins offered chapters which echoed my own pioneering ideas about 2012 and ancient Maya precessional astronomy. But of course my work was not mentioned, despite many communications with Carlson going back to 1994 and communications with Coggins in 1996. I wrote a query to the editors, Susan Milbrath and Anne Dowd, which I'll cover in detail in Chapter 4. I also wrote an email to Krupp — our first communiqué in over sixteen years, to ask him something specific:

Dear Ed Krupp,   June 8, 2015

As I recall, we last communicated in 1998 or 1999. I read with interest the recent anthology *Cosmology, Calendars, and Horizon-Based Astronomy in Ancient Mesoamerica* (eds Milbrath and Dowd). I was hoping you could clarify your comments in your preface. I note that you cited Aveni's 2009 book and mentioned the Xultun murals. Fascinating stuff! You also mentioned the 2012 "End Times Follies."

In your recent article of 2014, ("Archaeoastronomical Concepts in Popular Culture" in *Handbook of Archaeoastronomy and Ethnoastronomy*) you critiqued my work within the context of this "End Times Follies" (p. 278). In the more limited space of your recent preface you didn't mention me or my work, and I'm wondering if you still hold this association between my work and what you refer to as the End Times Follies. In addition, in another of your previous articles published in late 2012 (an updated version of your *Sky & Telescope* article, in *iQ Magazine*, Vol. 1 No. 5, December 2012, First Citizens Investment Services, titled "Time's Up: 2012 and the Maya Calendar End Times Follies") you depicted and critiqued my 1998 book, and also clearly portrayed me as a primary choreographer of the said Follies.

But just so I can be clear as to what your intended allusion in your recent preface was, I'd appreciate your quick response. Perhaps something in your thinking has changed. If space had allowed would my work also have been mentioned there, as being part of the End Times Follies?
        Sincerely,
        John Major Jenkins
        The Center for 2012 Studies - the2012story@gmail.com

I was wanting to confirm, directly from him, that his "2012 Maya Calendar End Times Follies" construct, which he alluded to in his Preface in the new anthology of 2015, included me. He did confirm this in his next email, stating that if he wrote another more detailed piece on 2012, "it would be folly to omit you." So, as I suspected, although I wasn't explicitly named in his Preface, his allusion to the End Times Follies was shorthand. His earlier article of 2014, in the *Handbook of Archaeoastronomy and Ethnoastronomy*, makes this clear.

Having re-established contact with Krupp, I now broached the subject of his flawed statements about my work. On June 10 I wrote:

I'd like to return to your comments in your first and second emails. I'm glad you acknowledge that I *haven't* argued that the Maya predicted apocalypse in 2012. You quoted me writing about "a tremendous transformation and opportunity for spiritual growth, a transition from one world to another."[27] The operative word here is "opportunity". My interpretation is that the Maya did not expect a sudden, predetermined "thing" or "prophecy" to "happen" on the period-ending date in 2012. The doctrine, or ideology, I identified and fine-tuned explicating, is that the Maya believed that, at period-endings like 2012, "deity sacrifice is necessary for world-renewal." And yes, it's a transformational process. For the ceremonialists engaging the deity sacrifice rituals, it can be experienced as a spiritual awakening or "spiritual growth".

There is a very important term missing in the quote from my work you provided. The accurate quote is: "*a time of tremendous transformation and opportunity for spiritual growth, a transition from one World Age to another*" (*MC2012*, 1998:XLI). (BTW, in 2011 Sven Gronemeyer characterized 2012 as an "era transition"; and Barb MacLeod described it as a "great return" of a cycle.) Notice that in my correct quote the transition is from one **World Age** to another. I realized early on in my work that a distinction had to clearly be made in the terminology used to discuss 2012 — the difference between "end of the world" claims and reconstructing an "end of a World Age" doctrine in Maya thought. In fact, this distinction is clearly discussed in the pages prior to my quote given above.

However, my discussion from the Intro to my book *Maya Cosmogenesis 2012* is precisely where you, in your Beckman Center presentation of November 2009 (which was done concurrently with the release of your *Sky & Telescope* article) have offered a misleading presentation about my work — one that definitely conveys the sense that I adopted and advocate an "end of the world" interpretation of 2012. Let's take a look. Here is a quote from your Beckman Center talk:

---

[27] In Richard G. Kyle's book *Apocalyptic Fever* (2012), he uses this inaccurate quote ("end of world" vs "end of World Age"), yet cites it to my 1998 book. Clearly, he took it, secondarily, from Krupp's "Great Scare" article. This demonstrates how the distortions of trusted scholars get repeated and disseminated.

"In this book [*Maya Cosmogenesis 2012*] Jenkins, in restating an unfounded belief, asserts **'the Maya believed the world will end in 2012.'** You will find that passage in that book. The Maya *didn't* believe that." — Ed Krupp, "Time's Up," The Beckman Center, November 2009. Mark 49:40. It's important to hear the inflected emphasis and tone of voice in your actual delivery of this passage: MP3: http://alignment2012.com/krupp1.mp3

You basically took a partial-sentence truncated passage out of context and conveyed a false notion about my work. The word "end" in the bolded quote above *is in quotes in my book* (p. XXXIV), which of course didn't get conveyed in your voicing of the line. This is the visual clue to the reader that the "end" is a provisional or questionable attribute given by whoever believes it to be true, like air quotes implies. And the "whoever believes it" is clearly referential *to the two authors I was just previously discussing*, where I said there was uncertainty or "doubt" about their sunspot flare theory. But "one thing is for certain" (this was my very next sentence, note the counterpoint between "doubt" and "certain"; I'm still characterizing their dire theory here): "The Maya believe the world would end in 2012." This was my sardonic *paraphrase* of *their* belief. You should appreciate and not a have a problem with this kind of theatrical mimicking of a dubious belief, because you've employed it yourself. For example, in your *Sky & Telescope* article you wrote:

"The ancient Maya of Mexico and Guatemala kept a calendar that is about to roll up the red carpet of time, swing the solar system into transcendental alignment with the heart of the Milky Way, and turn Earth into a bowling pin for a rogue planet heading down our alley for a strike." — Ed Krupp, "The Great Doomsday Scare," November 2009.

Aha! *You will find that passage in your article.* The very next sentence in my book, however, after your selective partial quote, should have given you pause: "But what does this mean?" I then go on to make the distinction between "end of the world" and "end of World Age," and identify my position (in reconstructing Maya beliefs) as belonging to the metaphorical application of the phrase (that it's an era birthing, ultimately about transformation and renewal). This is made clear throughout the five sentences that follow your extraction of part of a sentence, forcing it out context and misleading your audiences into getting a totally pejorative and wrong take on my work. The full passage (beginning after my statement that "I concluded that some doubt hung over the sunspot hypothesis" [of Cotterel and Gilbert]) reveals *the exact opposite* of what you conveyed to your audience:

"One thing is for certain: The Maya believed the world will "end" in A.D. 2012. But what does this mean? The end-times doctrine can be interpreted in two ways: metaphorically and literally. **My metaphorical interpretation is that the Maya believed that around the year we call 2012, a large chapter in human history will be coming to an end.** All the values and assumptions of the previous World Age will expire, and a new phase of human growth will commence. Ultimately, **I believe the Maya understood this to be a natural process, in which new life follows a death.**" [This is, essentially, an early languaging of my interpretation that, in Maya thought, sacrifice and renewal, death and life, go together and are both present at a 13-Baktun period-ending; the end of one World Age and the beginning of another.]

As a litmus test for how listeners would likely receive your portrayal of my work, I sent the mp3 clip of your Beckman Center talk to some random acquaintances, and every single one got the impression you were arguing that my work advocated doomsday. If that wasn't enough, you also stated in your talk that:

"Jenkins is probably more responsible than anybody for the current flurry and character of the interest in 2012 – although you gotta hand it to Argüelles for really lighting the fuse." Audio clip:
(mp3 http://alignment2012.com/krupp2.mp3).

This clearly refers, in the context of the 2012 doomsday movie just coming out (when you were giving your talk) and your overall portrayal of the 2012 topic, to the ubiquitous doomsday-2012 meme. And "lighting the fuse" is evocative of triggering a bomb going off, causing great destruction. Yes, others have hijacked my ideas for their own dubious theories and doomsday devices, and I've tirelessly spoken out against these misinterpretations. The media and documentary production houses have violated contracts and micro-edited my interviews to suit their purposes and distort my work. You don't adequately distinguish what my own stated intentions and findings have been from the unfortunate hijacking of my work by doomsday pimps and others. Like many critics of the "2012 phenomenon", you never mention that I've been critiquing the Maya calendar / 2012 pop marketplace for longer and more tirelessly than most. The very phrase "the 2012 phenomenon" was being used by Geoff Stray and myself long before it was appropriated by academic 2012 critics. (See my review-essay in *Zeitschrift für Anomalistik*, 2014: http://update2012.com/Jenkins-Zeitschrift-fur-Anomalistik-1-2014.pdf.)

You are basically blaming me (making me "responsible", see quote above) for the 2012 doomsday mess — for what exploitative fools did with my work. This is why I believe you perceive me as a primary choreographer in your "2012 Maya Calendar End Times Follies." The problem with your interpretation is that your are employing undiscerning guilt-by-association constructs. In other words, you don't acknowledge how I've written extensively against the doomsday mess, as well as ridiculous ideas such as we find in Argüelles, Calleman, and others. I have even long disagreed with McKenna's "sudden change" idea (see my discussion in *The 2012 Story*).

In your narrative, my work is a smooth continuation of early and quite dubious ideas about 2012 proffered by others. And you don't acknowledge a fundamental difference between my 2012 work (beginning in earnest circa 1993) and what came before. Which is that I took 2012 seriously as a legitimate topic of rational inquiry, regarding what the ancient Maya thought about it. Many scholars, including perhaps yourself, have been unwilling to entertain that rational investigators might be able to reconstruct what the ancient Maya thought about the 2012 period-ending date. I was doing this as an extended investigation at a time when almost all scholars and academic publishers treated 2012 with derision. And many still do, *despite the two 2012 inscriptions we now have at hand.* My approach was to examine the pre-Classic culture, and site, that informed scholars had already credited with formulating the Long Count calendar. My logical and reasonable approach to the topic thus followed, with striking and potentially revolutionary results (revolutionary in terms of how we understand the level of astronomical knowledge among the pre-Classic people of Mesoamerica, and the sophistication of their integrative religious ideologies).

Thus, much of my work springs from my interdisciplinary analysis of the Izapa civilization. My two-part interpretation is simply stated: **the creators of the Long Count intended the 2012 period-ending date to target a rare astronomical alignment within the cycle of the precession of the equinoxes, and saw this alignment as signaling (not definitively *causing*) the need for deity sacrifice in order to facilitate worldrenewal.** There it is in one sentence, less than 50 words. This is, in essence, what I reconstructed at Izapa and argued in the mid-1990s, while also drawing from other sites and evidence elsewhere. I've been happy to fine-tune this definition of my work over the years, but the ideas are there, in *Maya Cosmogenesis 2012.* And sure, some of the languaging in that book is not as clearly expressed as I would now prefer, but that's the nature of evolving writing skills and honing the clarity with which ideas are expressed, over time. And there is a lot of post-1998 writings and research, including my 2010 SAA presentation (http://www.thecenterfor2012studies.com/MEC-Facebook-Discussion-2010-ON-Jenkins-SAA-TRT-Astronomy.pdf), that critics have overlooked. For example, your later articles and presentations don't cite my 2009 book, *The 2012 Story,* which came out the month before your *Sky & Telescope* article and provided exhaustive critiques and discussions of the 2012 milieu. That book sold widely and internationally and was translated into several languages, in paperback and hardback. I suspect that at least some critics ignore that book because I also took to task Maya scholars (someone had to do it) for their ridiculous and fallacious assertions about 2012 (notably, Aveni at Tulane).

Now, I'd like to point something out. I'm not trying to convince you of my findings. However, what is striking is that other Maya scholars have been recently echoing my own two-part interpretation, which I originated some two decades ago. For example, in the same anthology that you wrote your preface for (*Cosmology, Calendars, and Horizon-Based Astronomy,* 2015), John Carlson offers an interpretation for how the ancient Maya thought about 2012. He writes that it involves *deity sacrifice as a necessary prelude to world-renewal.* Clemency Coggins advocates for a pre-Classic precession awareness, and explicitly reiterates my model for how the cosmic center deity evolved as human groups migrated into the Tropics (*MC2012,* 1998, 31-40). She also echoes certain ideas I've reconstructed about Izapa, but she neglected to note the Izapa ballcourt's alignment to the December solstice sunrise. (In addition, elsewhere Michael Grofe is finding that the ancient Maya could accurately track the Sidereal Year and the Tropical Year — supporting evidence for my 2012 "galactic alignment" reconstruction. See his peer-reviewed essays posted on his Academia.edu page). So, as part of your Follies narrative you would likely have critiqued and dismissed my ideas in your Preface, if space had allowed (your citation to Aveni in any case serves this purpose), but meanwhile two other scholars *in the same anthology* echo key ideas in my two-part 2012 interpretation — in both the astronomical and ideological aspects. (Not to mention that curious book jacket image, which seems to depict a pre-dawn version of the sun's alignment with the nuclear bulge of the galaxy, a.k.a. the galactic alignment.)

So, how do you reconcile these facts:

- In the mid-1990s I emphasized that my work was about reconstructing what the ancient Maya believed about 2012. I found evidence and an astronomical key that was symbolized in Maya Creation Myth concepts and other evidence, and this was a departure from what came before in the work of others. (Argüelles's "galactic synchronization" is NOT at all the same as the precession-based galactic alignment. McKenna's definition of it was quite loose, and he didn't explicitly connect it to a Maya intention with 2012 in his 1975 book (he gave it a 200-year range)).
- A simple definition of what my findings are is available in my many articles, presentations, web pages, interviews and books, but has never been accurately portrayed by my critics, despite direct communications with them over twenty years. More usually, my work is conflated through guilt-by-association allusions with doomsday theories or other writers in the marketplace.
- My two-part interpretation (a precessional alignment of the solstice sun with the Milky Way/ecliptic Crossroads & southern terminus of the Dark Rift that signals the time for deity sacrifice in order to facilitate world-renewal) is now being echoed in the statements about 2012 published by Maya scholars (not as a coherent integration, but elements are found in Coggins, Carlson, Callaway, MacLeod, Grofe, Dowd, Barrientos, Gronemeyer, and others).

That third point is a far cry from where Maya Studies stood on 2012 when I first published my pioneering 2012 work in the mid-1990s. Acknowledgment or cognitive dissonance? Well, mitigating comments by critics continue to be published. Your own allusions to the 2012 Maya Calendar End Times Follies indicts my work as dubiously part of the same marketplace mess that I've actually critiqued and exposed the fallacies of. And you do, in your preface, cite Aveni's 2009 book as support for your dismissive attitude. My critique of Aveni's book will open up another can of worms, perhaps best reserved for a separate discussion. Do you believe that Aveni's critiques of my ideas and position, in his 2009 book, are accurate and reliable? It seems you do.

I should close here, and this will have to serve as an overview. It's very difficult to offer a brief recapitulation of the issues I've seen with your critiques of my work. Some are just purely factual corrections (can send if you like) but many are conceptual, in terms of your forced narrative of where my efforts fit within the evolving 2012 discussion, my presumed "influences" and associations. There is increasing cognitive dissonance within Maya Studies as the ideas I've long articulated with good evidence and argument (but which are rarely accurately summarized or even acknowledged) are slowly being subsumed into the general consensus. This is, of course, the three stages by which breakthroughs in a field of study are integrated: 1) the originator of breakthrough work is ignored, 2) the ideas and the originator are vehemently attacked, and 3) the breakthrough ideas are accepted as if they were known all along. Sincerely,

John Major Jenkins
The Center for 2012 Studies - http://thecenterfor2012studies.com
Attachments here, including: For-EdKrupp-6-10-2015.doc

In response to this email, Krupp once again (like 16 years earlier) vaguely asserted that he was busy and would try to respond later, also asserting that he "didn't agree." I awaited his response, but there was none. So I prepared the promised package of books and article off-prints and, much like years before, sent off my work to Krupp. I mailed it a month after the email above, with the following cover letter:

Griffith Observatory
Los Angeles

Dear Ed Krupp,   July 9, 2015

I am enclosing a booklet and a selection of articles and chapters. It strikes me that your comments and critiques about my work are not very well informed regarding what I've actually written through the years. Much hinges on your distorted, partial, out-of-context extraction from the Introduction to my 1998 book. You don't seem to have read or considered my lengthy letter to you of early 1999, after I sent you my book for review. That was the last I heard from you until I reached out last month (June 2015). You easily could have contacted me at any point during the interim period of 16 years, and I would have been happy to dispel your mistaken notions about my work and clarify any questions. Your updated *Sky & Telescope* article for the *IQ Magazine* (December 2012), would have benefited from a reading of my 2009 book *The 2012 Story*, which you claim to have on hand. Likewise, the same issue applies to your *Handbook for Archaeoastronomy and Ethnoastronomy* article of 2014.

A few basic points about my work that my critics prefer to ignore: I investigated 2012 as a valid artifact of ancient Maya thought. My approach has always been *to try to reconstruct what the ancient Maya thought of 2012*. I asked the right questions, regarding the origins of the Long Count (the "2012" calendar), and studied the most relevant time and place: Izapa and the Izapan (Isthmian) civilization. I investigated evidence for how the creators of the Long

Count thought about 2012, during a time when all Maya scholars treated 2012 with contempt and derision or, at best, something to be examined only as a psycho-social phenomenon of the modern marketplace. In August of 2010, the first serious treatment of 2012 by so-called "real" scholars, as a valid artifact of ancient Maya thought, was published (MacLeod and Gronemeyer, Wayeb no. 34). This occurred four months *after* my SAA presentation on the Tortuguero inscriptions.

Thereafter followed the Oxford IX conference in Peru, with the academic papers on 2012 being published by the IAU in July 2011. Then, the Gelfer anthology (in which I contributed chapter). Then *Archaeoastronomy Journal* Vol. 24 (August 2012). The papers in these journals which suggested, proposed, deduced, or reconstruction anything about what 2012 might have meant to the ancient Maya included the core ideas that I published in the mid-1990s: the precession-based galactic alignment (properly understood), and a doctrine of period-ending deity sacrifice & world-renewal. Please read the essays by Carlson, Grofe, MacLeod, and Callaway if you are doubtful. And now we have Carlson's 2015 essay and Coggins' essay, in the anthology you wrote your preface for — both of which echo my long-ego articulated ideas. Comments? If you don't understand that my work was centered on an interpretation of deity sacrifice & world-renewal which the ancient Maya saw timed by the galactic alignment, you should re-read *Maya Cosmogenesis 2012*, the core ideas of which are summarized in a concise section of my 2002 book *Galactic Alignment*.

I enclose the following:

My booklet of late 2012, titled *Reconstructing Ancient Maya Astronomy*, based on my presentation at the New England Antiquities Research Association (NEARA).

"Approaching 2012: Modern Misconceptions versus Reconstructing Ancient Maya Perspectives" (2011, in the anthology edited my Joseph Gelfer, preface by Michael Coe).

"The Coining of the Realm (of the 2012 Phenomenon)" in *Zeitschrift für Anomalistik* (January 2014). This is my critique-review of an essay by John Hoopes & Kevin Whitesides (2012), in the same journal.

"12/21/2012: The Game Ball Goes Through the Goal Ring," *Institute of Maya Studies Explorer*, Vol. 41, Issue 12. (answering questions)

"Fear and Lying in 2012-Land," January 2009. Critique of the 2012 phenomenon, in the anthology *You Are Still Being Lied To*, ed. Russ Kick.

I also enclose a print-out of the email I sent you on June 10, which contained comments and questions regarding your portrayal of my work. That was four weeks ago, and you haven't responded. I think a responsible scholar should respond to the person whose work he was maligned and inaccurately assessed.

To this end, please see the Open Letter piece, also enclosed. This contains many of the points from my email to you of June. It also indicates factual errors in your various articles on 2012. There are many more, but I've selected a few. **I invite you to respond to this Open Letter which corrects and critiques your statements about my work.** It will be posted at *The Center for 2012 Studies* (http://thecenterfor2012studies.com); and I will be happy to post your rebuttal when you provide it. My email: the2012story@gmail.com.

The skewed narrative and basic factual errors in your treatments obviously suggest that you aren't that well versed in the ideas and publications of "the 2012 phenomenon," which I've been critiquing since the late 1980s. Read, for example, my "Fear and Lying in 2012-Land" piece or the several in-depth chapters in my 2009 book *The 2012 Story*, which was published before your *Sky & Telescope* article. Early critiques (Argüelles, etc) are in my 1992/1994 book called *Tzolkin* (Borderland Sciences Research Foundation).

It's funny, in 1996-1997 I was inquiring at academic publishers regarding publishing my magnum opus, *Maya Cosmogenesis 2012*. The conversation was over as soon as I mentioned 2012 — within 45 seconds of the phone call's initiation. Why? Because, of course, 2012 was not a valid topic of rational inquiry. For scholars and for academic publishers, that's been the case until a smattering of serious scholars took it up, beginning in 2010. Which is just to say that my work was too far ahead of the curve to be of interest to academic publishers in 1996. I don't think you've read the academic literature on Maya precession calculations, the 3-11 Pik formula, or Tortuguero Monument 6. If you had, you would see how my interpretations of what 2012 meant to the ancient Maya started being echoed much later by scholars. But many of them have gotten the memo jammed down their throats, from Aveni or Stuart, that 2012 cannot be treated seriously! Don't mention or cite JMJ (Carlson, Aveni); don't explore the unique ways the Maya tracked precession; and any talk of world-renewal is New Age rubbish (Hoopes).

Finally, I did not print out my *Society for American Archaeology* paper "The Astronomy of Tortuguero Monument 6." In 2009, I was invited by Drs Robert Benfer and Larry Adkins to speak on the SAA panel "Archaeoastronomy in the Americas," which happened April 15, 2010. My read paper was included in the 206-page debate about it, sponsored by Dr. Ed Barnhart and the board members at the *Maya Exploration Center*. This occurred in late 2010, with comments by Stan Guenter, Gerardo Aldana, Michael Grofe, Barbara MacLeod, Ed Barnhart, and others. The entire transcript, with my concluding comments, is posted at Barnhart's MEC website (http://www.mayaexploration.org) — and it's also at *The Center for 2012 Studies* website. Please read my SAA paper in that file. In the larger debate you can perceive the juvenile and irrational debate & critique / evade & repeat tactics of anti-2012 scholars like Guenter and Aldana. They project upon me something that I am not, which is akin to the worst kind of bigoted, mean-spirited, scape-goating. Really amazing to experience, and disappointing to see professional scholars behaving so badly.

I hope you will file my enclosed booklet and articles in your Griffith Observatory library, so you and your colleagues can be accurately informed about my work. The skewed, under-informed, juvenile, and unprofessional critiques of my work are documented for the record. I also hope that whatever future attempts you make to critique my work will be informed by these essays, and take into account the odd circumstance that the two core ideas about 2012 that I articulated in the 1990s were echoed years later by other Maya scholars — notably by John B. Carlson in the same anthology of 2015 (*Cosmology, and Calendars, and Horizon-Based Astronomy…*) that you wrote a preface for. Yep, the ancient Maya believed that, at the period-ending of 2012, deity sacrifice is necessary for world-renewal. Best wishes,

John Major Jenkins • The Center for 2012 Studies

A few weeks later I emailed him to ask if he'd received the package, and he briefly confirmed. My approach to Krupp could not easily appeal to a higher authority which might objectively assess his behavior, as I'd done with Morrison and Aveni. *Sky & Telescope* is a private publication notorious for abetting diehard "skeptics" who will try to deconstruct anything, and who are not in fact true skeptics, but pseudo-skeptics adamantly fixated on disproving what threatens their personal convictions, mainly those tied to atheism and scientific materialism. They are not unlike fundamentalist believers who launch ideological jihads against perceived foes. There would be no chance of a fair hearing there, or even in getting a chance to publish my rebuttal to Krupp's article. The fairness doctrine in journalist was trashed years ago, and mafia-style scientist-terrorists rarely allow their victims to defend themselves in an open, fair, and unmolested way.

Likewise, the Beckman Center is a private presentation space which nevertheless gets funded by various sponsoring agencies. In Krupp's case, his talk was sponsored by the AIA (the Archaeological Institute of America, which publishes *Archaeology Magazine*). So, there'd be some chance of appealing to their policy of academic accountability, but not much. I decided to take the more direct route and appeal to Krupp's sense of honesty in responding to my questions and comments. Sadly, what I'm seeing here is a replay of what he did sixteen years ago. Which is to *not respond* directly to me, but to bide his time, muttering about his busy schedule, while he plans to eventually launch an offensive indirectly and then run away again when detected. His ammunition is taking truncated quotes out of context and cleverly using loaded lingo and guilt-by-association insinuations. Not to mention spewing a cart-load of factually incorrect things, as I detailed in my official Complaint to him (in full, in Appendix 3). This is more the behavior of a propaganda minister than a reputable astronomer practicing sound scholarship.

His only face-saving response when confronted with these corrective facts and uncomfortable questions that expose his malicious intentions is to shut up and walk away. See ya in sixteen years, Ed.

◇ ◇ ◇ ◇ ◇

John Hoopes: The Wolf in Sheep's Clothing

John Hoopes, an anthropology teacher at Kansas University, has maintained a pejorative and utterly misleading attitude to 2012 for about a decade, despite new evidence and work done by his colleagues. Dealing with his under-informed attempts to discredit myself and others has been likened, by one researcher, to dealing with "a weasel on steroids." These kinds of characterizations aside, I'm willing and able to assess his critiques and comments on their own merit. It's taken a little while to understand where he's coming from and what his strategy is, because he first approaches you in a conciliatory, friendly way, asking to better understand your work. He asks some general leading questions; you answer the best you can. Then, months or years later, you find that he's selectively cherry picked things you stated and cleverly twisted ones perceived "associations" with other writers. Between early 2007 and early 2011, when I was actively and openly communicating with Hoopes, I gave him the benefit of the doubt and endlessly tried to help him understand what my work was about, where the evidence was, the academic sources and methods I cited and used, and so on. Very little of that ever factored into his so-called "critiques," which are more akin to pejorative constructs, invented "influences" and guilt-by-association assertions, all intended to mitigate.

Hoopes's efforts are intimately bound together with the position of his close friend, John B. Carlson, who is the Editor-in-Chief of the *Archaeoastronomy Journal* (AJ), published since 1999 with the University of Texas Press. Although I'll discuss Carlson with more focus in the next section, his role in Hoopes's first 2012 publication can be mentioned here — because Carlson approved, published, and defended it. In the Volume 22 edition of the AJ, which was released in early 2011, Hoopes published a review of Van Stone's and Aveni's books on 2012. In it, he snuck in some comments on my work, which asserted and crafted several false and misleading things about my professional background, my previous employment, and my work in Maya astronomy. In fact, Hoopes was not able to even acknowledge that my work, as I myself have stated in print and to him directly in personal emails, is about Maya astronomy. Instead, Hoopes crafted a false framework that the galactic alignment (an alignment within the precession of the equinoxes) is essentially, first and foremost, *astrology*. This is as absurd as asserting that a sunrise or a full-moon is, essentially, at the most basic core level, astrology.

But his construct had a point, for in doing so he could then assert that my work was rooted in astrology, a pseudocience with no scientific merit. He bolstered his false construct by asserting that I once worked as a professional astrologer (totally false, never even tried). My earliest article on the galactic alignment (1994) was published, yes, in a magazine called *Mountain Astrologer* (for reasons of local convenience as I've previous noted), but I explicitly distinguished my discussion as being about *the precession of the equinoxes* and an *astronomical* alignment, which I illustrated with several EZCosmos astronomy program diagrams. To further misrepresent and denigrate my background and work, he asserted that my galactic ideas (i.e., my galactic alignment reconstruction) were inspired by the writings of astrologer Dane Rudhyar. This is totally false; Dane Rudhyar never talked about the galactic alignment. I've never cited Rudhyar for any prior articulation of my galactic alignment work, which I've identified as unprecedented and pioneering, so Hoopes's statement is tantamount to an accusation that I plagiarized my work from Rudhyar. Hoopes had crafted this assassination of my character and my work out of the delusional fictions of his imagination, designed for defamation.

In addition, Hoopes didn't even try to support his assertions with citations to anything. They are unsupported assertions, a fable. This is clearly a textbook example of Poor Scholarship 101, yet it was published in a reputable peer-review journal published under the auspices of the University of Texas Press. So, how did it pass through the filters of peer-review and editorial scrutiny? The editor, John B Carlson, did not flag these statements, although such inflammatory and unsupported assertions about a living author would have been immediately flagged by any number of discerning non-academic trade publishers. Defamation is a violation inside and outside of academic publishing.

Well, this occurred in early 2011. I was still unused to such vile behavior by scholars, and felt that it should be a simple matter to clear it up. I emailed Hoopes and asked for his sources that would support his statements. After an initial acknowledgment of contact, before he knew I was going to ask for his sources on his slanderous statements, he stopped responding to my queries. I then contacted him on his Facebook account. He blocked me. So, in September of 2011 I brought the situation to the attention of Sue Hausman, the journals manager at the UT Press. She asked me to provide the details, which I did:

Dear Sue Hausman,

Thank you for your attention to this issue. The PDF of John Hoopes' review in Vol XXII of Archaeoastronomy was freely posted on Mark Van Stone's website, and that is where I accessed it. The statements in question are found in the right column of page 143:

"The "2012 Phenomenon" makes much more sense in the context of astrology than astronomy, as becomes clear from the influence of astrologer Dane Rudyar on New Age prophet and 2012 guru José Argüelles and on John Major Jenkins (who once worked as a professional astrologer) ..."

(further down the column):

"His [Rudyar's] book *The Planetarization of Consciousness* (1970) helped inspire the first Whole Earth Festival while *The Sun is Also a Star* (1975) provided the intellectual underpinnings for claims by Argüelles (for whom Rudyar was a personal mentor) and Jenkins about ancient Maya concerns with the movements of the Sun relative to the Milky Way galaxy.   ... astrology is a pseudoscientific "fringe" discipline."

I am not a professional astrologer, never have been and never tried to work as one. An early book of mine ([*Tzolkin*] 1992) criticized pop / causal astrology. Hoopes's intent to defame is evident in the (false) identification of me as a professional astrologer, in the misleading association of my astronomical reconstruction work with an astrological context, and with the assertion that astrology is pseudoscience. I informed Hoopes by email quite some time ago that I was only vaguely familiar with the name Rudyar. Having subsequently looked into Rudyar's writings, I find that they have nothing to do with my reconstruction of precessional astronomy in ancient Mesoamerica, nor the arguments and evidence I've brought to bear on my thesis — accept for the shared use of the term "galactic." Since I know my work to be, and present it as being, unprecedented, and I don't credit Rudyar with it, Hoopes's statement is tantamount to an accusation of plagiarism. These are very serious lapses in scholarly professionalism, accountability, and ethics. It's unfortunate that such comments were not flagged for checking, and that they've already appeared in print. They are totally false, designed for defamation. Even the trade publishers I've worked will flag questionable comments for checking, as a standard procedure. I've tried to seek a response from Hoopes, or an explanation, but he has refused to respond.

My suggested solution:

    1. Facilitating a response from Hoopes
    2. A printed correction in a future edition

A possible future problem must also be addressed. As I mentioned on the phone, my additional concern is that Hoopes's under-informed and incorrect statements will appear in Hoopes's forthcoming essay in the next *Archaeoastronomy journal*, which features papers on 2012 by MacLeod, Grofe, Callaway, Carlson, and other presenters from the 2011 Oxford IX Archaeoastronomy conference in Peru. Since there is such a highly politicized climate around the 2012 topic, and much misinformation about my work and ideas, I would prefer that I would be allowed to vet for accuracy anything that was written about me and my work in the pages of *Archaeoastronomy journal*. Thank you for your time. Best wishes,

    John Major Jenkins
    kahib@ix.netcom.com

I had expressed, as can be seen, a concern that Hoopes would continue to assert false and slanderous things about me and my work in Volume 24 of the same journal, which I knew was in the works with a contribution from Hoopes. It was to be the "2012" issue of *AJ* with contributions from Carlson, Hoopes, Van Stone, and other scholars who were finally taking a closer look at 2012. A few of them were even making an attempt to reconstruct what the ancient Maya thought about 2012. So, it would be a landmark publication and Carlson would, of course, be the editor.

I offered to vet Hoopes's article in order to avoid further problems for the UT Press.

Hausman handed my requests and emails over to Carlson, who responded by requesting a copy of my 1992 that I'd mentioned in my email. Apart from the fact that I knew he already had a copy of this book, his email was evasive, and was an attempt to turn the tables and make the issue about me defending my work. This was NOT the issue, as my next email to Carlson emphasized:

Dear John Carlson and Sue Hausman,

To clarify, I am simply requesting that John Hoopes supply proof for his statements, published in Archaeoastronomy journal, Vol. XXII. I was hoping that Sue Hausman could facilitate a response from him, since he refused to respond to my several email requests last month. In this regard, his statements are explicitly asserted but no citations or proof were provided for them. That is what I am requesting be provided. The inability of Dr. Hoopes to provide such evidence for his statements will determine whether or not my "accusations" are warranted. Please re-read the details of my email below, lest I be forced to repeat myself.

Your desire to assess a previous 1992 publication of mine is irrelevant to this request. This inquiry should rather be directed to Dr. Hoopes, who bears the onus of providing proof, citations, or some kind of evidence for his statements. I am offering the benefit of the doubt, and would appreciate a straight forward response. Whatever role the editors of the journal played in allowing his unsupported statements to make it through to publication, without proper flagging and professional fact-checking, is a different matter.

Sue, your attention to resolving this matter will be greatly appreciated. For your convenience my earlier email is copied below [see p. 86]. Sincerely,

> John Major Jenkins
> Cc to Sue Hausman

And again, after another evasive response from Carlson:

Sept. 19, 2011
Dear John Carlson,

You wrote that you "see nothing in what Professor Hoopes wrote in his review that is factually incorrect." How do you know what he wrote is not "factually incorrect" if he did not supply the citation for what he stated as fact? How are readers going to know, or verify for themselves, if he did not supply a source for what he said, which amounts to defamation in the context of his overall treatment? Let's review:

Step 1: My request asks John Hoopes to supply the citation(s) to the sources that support what he asserted as fact. These were defamatory statements that you, as editor-in-chief of UT's Archaeoastronomy journal, should have flagged for fact-checking. Your failure, as editor, to do so is a fact of the matter.

Despite your bewilderment at my refusal to cater to your irrelevant evasions, Step 1 is what is required before any other matters need be pursued. I have indeed been trying to take this up with Dr Hoopes, as you suggested, but he has refused to respond to my queries. This is why I was asking the journals manager, Sue Hausman, to help in facilitating a response and resolution to the situation.

My question to you is: as editor-in-chief, do you not have a policy for flagging and fact-checking unsupported statements, especially if those statements amount to defamation of a living author in the context of the associations asserted in the construct of the piece? I feel this is a very serious breach of professional ethics. As a matter of decency, I will reiterate my concern for what Hoopes may be writing in his contribution to the forthcoming issue of *Archaeoastronomy*, which contains expanded essays by the contributors to the recent Cambridge IAU 278 journal. In the interest of clarity, and accurate fact-based presentation, I have offered to review and fact check his article. I had assumed this would be a welcome invitation in order to preserve the reputation of your journal. Thus, the matter is not ended here, as you would wish, but will continue if Hoopes's unprofessional and sub-standard scholarship — his tactic of baseless character assassination and defamation — continues to be sanctioned and/or overlooked by you, John Carlson, the editor-in-chief of *Archaeoastronomy* journal. Sincerely,

John Major Jenkins

And yet again, after no response:

I feel it is necessary to put a fine point on my inquiry. I am not making "a case" to you. I am not inviting you to receive a defense of my work, or assess evidence from me regarding my work or anything that I have written or believe. The issue is with your author, John Hoopes, asserting as fact things about me which are not supported with evidence or citations in his Archaeoastronomy Vol XXII review/article. This fourth email reiterates that I am inviting a clear response to a simple question: can Hoopes supply the evidence or citations for the statements he made? (For the record, my effort here follows several emails to Hoopes several weeks ago which went unanswered after his initial acknowledgement of receipt.) Can you facilitate a response from John Hoopes regarding this issue? If so, please relay his response to me. If not, please explain why. Sincerely,

John Major Jenkins

In all of this, Hoopes was nowhere to be seen. Hausman, caught in the middle and happy to pass it off on Carlson, faded into the background. When the communication with Carlson did not resolve in any kind of honest way, I appealed again to Hausman, but she dodged. A few months later I tried again but someone in the Journals office at UT Press said I should — if I caught the challenge correctly — file a legal claim. I was not expecting or wanting to take it to that level, because I thought the problem would be addressed by honest, ethical scholars and their honest, ethical university press publisher. You know, just an errata in a future edition of the AJ journal. Not a lot to ask, but apparently impossible to achieve.

Something else seemed to be at play here, and I soon got a clue. It involves Carlson's interpretation of 2012, and how closely his idea reflects my own. I'll reserve this for the section on Carlson. For now, continuing with Hoopes, I should say that I'd already critiqued his "Mayanism" construct, which he'd developed on Wikipedia since 2008. My 2009 book, *The 2012 Story*, treated his work fairly but had to point out certain problems with it, namely, that he'd appropriated a term that other scholars had used, a few years before, in a way completely opposite to how he used it. Hoopes was basically crafting a pejorative container for unwanted ideas and outsiders who he cleverly maneuvered into his cage, through various guilt-by-association tactics. For example, he would craft ideological equivalencies between myself and other authors because we were published by the same "New Age" publisher, or spoke at the same conference. And he never could accurately — ever — portray or summarize what my work was actually about or give credit for the unprecedented ideas I've proposed. He resorted to superficial and misleading cardboard cut-out versions of my work. It went so far as to hear (from friends who attended his presentations) of him devising Blavatskian-Theosophical and Nazi precedents to my work. Ugly, very ugly stuff, indeed.[28]

His efforts came to a head with his submission, with Kevin Whitesides, to the German journal *Zeitschrift für Anomalistik*, which in early 2011 offered a prize for the best article on 2012. It went to the piece written by MacLeod and Van Stone (to be discussed in Appendix 1 online), but the Whiteside-Hoopes essay was also published in the same edition of that journal, appearing in mid-2012. I didn't see it until mid-2013. And when I did, I thought, okay, I've had enough of these shenanigans. This is really where my Experiment began. As always I began by giving the perps the benefit of the doubt, and asking them to clarify their statements and respond to the factual corrections that I could easily provide. So, in the summer of 2013 I contacted Hoopes and Whitesides by email. By this time Hoopes knew the jig was up for him, as I'd seen through his games and the issue with AJ Vol 22 had never been resolved. So, his tactic in responding to me was to ask me to provide copies of my books so he could truly assess my ideas. This had nothing to do with the matter under consideration. It was an evasion strangely reminiscent of Carlson's behavior, asking me for other irrelevant sources and turning the tables.

Whitesides, on the other hand, responded to a few of my questions, which clarified who had said what in their article. (Co-written articles are funny beasts — they provide shelter and confusion over which author is responsible for which malfeasance. It's best to consider both to be equally responsible parties.) Whitesides confessed to apparently getting his sources wrong, since one of the issues involved him citing my 1998 book for proof of his ridiculous critique, when

---

[28] I refer to the Austin conference of 2011, where Robert Sitler took Hoopes to task for what he was insinuating in reference to my work. Insinuation is all that is needed in the game of mitigation.

that book didn't supply any evidence for his hallucinated reading of my work. In fact, it provided evidence for an opposite view of my methodology.

Their article stated that I was engaging a "hermeneutic" in which I believed my ideas were not subject to scholarly assessment and critique — sort of like inspired visions or revelations of truth above mere rational analysis and debate. Incredible! I've been insistently placing my reconstruction work before the eyes of scholars for two decades, hoping for intelligent critique and discussion! Hoopes and Whitesides had both been included in my efforts on this front for many years. This was asserting the exact opposite of what I've done, like claiming that Abe Lincoln was a slave trader. As I speculated, it may be that Whitesides was confusing my original pioneering reconstruction work (in my 1998 book) with his misreading of a later elaboration in my evolving work, in which I discussed the value of a Perennial Philosophy based in universal principles and how certain spiritual teachings, including those in the Maya Creation Myth, were expressing such universal wisdom teachings. But this is *not the same* as claiming that my reconstruction of ancient Maya cosmology is a divine revelation not subject to rational critique and analysis. The fallacy of such a construct is so patently obvious, that it's hard to believe two PhD-holding scholars could commit such folly.

After the initial cordial exchange with Whitesides, he said he'd need a few months to get a considered response back to me, as he was busy with other things. I waited almost three months, and then contacted him again. He was now perturbed at my persistence, and it seems he must have been advised via conversations with Hoopes as to how to approach the problem. He evaded and dodged. By November, four months after my initial query, the writing was on the wall — I would receive no considered or rational response from either Hoopes or Whitesides. And so I initiated Plan B — contact the journal that published their baseless screed, and offer to write a rebuttal. Surely, journals have such opportunities in place. *Zeitschrift für Anomalistik* was based in Germany, and they cordially responded that, yes, I could write a review-essay rebuttal, but it needed to be brief and address only the points of error. So I did, following the restricting guidelines they provided.

By December 17 they had accepted my review-essay. Part of their policy allows the authors a chance to respond. They can accept or decline. So my piece, which had already been peer-reviewed by their editors and, after a few minor changes, accepted for publication, was forwarded to Whitesides and Hoopes. Things had happened fast since my last communiqué with Whitesides in mid-November, and they must have been surprised. Also, the Christmas break for teachers was upon us, and I was preparing to depart for a tour in Peru. Nevertheless, Hoopes took the lead and responded with his arsenal of evasive tactics. It was quite astounding to witness and experience, but not that surprising. The weasel on steroids went into overdrive.

First, as some pretense for him writing an informed response to my corrective review, he sent a request to me that I send him all of my multi-genre writings since the 1980s, based on a bibliography of my works I had posted on website. The list included poetry, family genealogy research and personal biography sketches, and experimental quasi-fiction — virtually my entire output since the mid-1980s. This was a totally irrelevant request as I had been careful, in my review-essay, to provide online links to the evidence from various previous exchanges and writings.

Hoopes even requested material that I knew he already had. I found this to be evasive and ridiculous, but I instantly knew that he was doing this as a pretext for claiming, later, that I refused to provide the necessary materials with which he could make a response. Hoopes fishes for useful bits of personal history, or fragments of quotes, that he can use to work up a compromising narrative. It is the only recourse of a desperate pseudo-scholar trying to cover up his bad behavior and, most likely, his jealousy issues. (Hoopes called me in late 2009, asking for advice on book publishing and the phone number of my editor at Penguin.) Every scholar wants their work to be read by millions and to make a breakthrough contribution, but few actually do. It is unacceptable to them when an outsider does this — one who didn't swallow goldfish at fraternity hazing parties (or worse) and who didn't go into debt, paying the piper, through years of boring classes.

With this as the likely scenario I decided to play his game and respond to Hoopes at length, to explain the nature of the various writings I'd done, sending various excerpts and links. Several of the books he listed weren't even sold in any large edition, and I literally had only one or two copies remaining, which I reserved for my personal archival library. I certainly wasn't going to ship boxes of my books, some of them rare, off to Hoopes. Nevertheless I cooperated as much as possible — and more than anyone might reasonably expect — which can be clearly seen in my first response of December 21, 2013, followed by additional responses in the following weeks (see details in Appendix 3 online). But Hoopes continued to demand all of my previous writings be sent to him, ignoring what I knew he already had many of them and that I'd already provided links (at least one of the books was online for free), jpg snapshots of pages in my books, and ordering info. Whitesides, for his part, took a backseat to these proceedings, although he was cc'd.

Before I left for Peru I thought it might be prudent to explain to the editor of the journal that Hoopes was making unreasonable demands. I felt Hoopes might be making a bid to derail the publication of my review-essay and, in the ensuing confusion this almost happened. The editor responded that Hoopes could use any sort of previous material he wanted. Well, that's certainly true, but I felt the editor misunderstood that Hoopes was asking that I send him my entire multi-genre output of 30+ years. The editor was probably put-off and disturbed by this needless mucking around — which is probably what Hoopes was shooting for. Hoopes was confusing the focus, was going off track, trying to make the issue about me defending my work, rather than him responding to the specific points and corrections I'd raised in my review, regarding factual errors and misrepresentations in the article he wrote with Whitesides. For that was a stated stipulation of the process, according to what the editor had previously sent me — that both parties must stick to the actual content of the items raised.

We all took a hiatus during the Christmas and New Year break. But on January 8, the day after I returned from Peru, Hoopes was ready with fresh ammo. A fiasco of desperation unfolded, which is all documented in the additional documentation links in Appendix 3. Finally, on January 11 the editor intervened and reaffirmed his intent to publish my review-essay, and Hoopes and Whitesides were given until the end of the month to complete their response, if they decided to offer one. And so it transpired, and then the long wait to publication ensued. I didn't see their response until both my review and their response was released on August 1, 2014. In my review-essay I had identified many unambiguous errors, including their claim that Sitler (2006) coined ("first used") and defined the phrase the "2012 phenomenon". This claim was easily proven false, and I supplied the evidence. However, incredibly, they managed to wiggle around admitting to this error, and *all the others I'd proven*. Their response was a textbook exercise in how to be a sleazy, intellectually dishonest, pseudo-scholar.

Here are some choice quotes from my review-essay. I begin my review with a correction to their incorrect belief as to who coined the phrase "the 2012 phenomenon":

In their article in *Zeitschrift für Anomalistik* (2012), Kevin Whitesides and John Hoopes state that Robert Sitler was "the first to use and define the term '2012 phenomenon'" (Whitesides and Hoopes, 2012:50). They cite Dr. Sitler's 2006 *Nova Religio* essay as the source. But Sitler himself, in his essay, mentions Geoff Stray's book *Beyond 2012*, published in 2005 and which Sitler states "promises to be the most comprehensive book on the 2012 subject to date" (Sitler, 2006:29). In that book of 2005, Stray uses the term "the 2012 phenomenon" more than once,[29] first in a note to his Introduction, where he states:

"www.diagnosis2012.co.uk – also known as 2012: Dire Gnosis, where *Dire* means serious or urgent, as well as dreadful, and thus sums up the ambiguous nature of the 2012 phenomenon" (Stray, 2005:288)

Stray's book was written largely in 2003, with the first manuscript completed by March 2004 at the latest (when he sent it to me). Sitler may have been aware of Stray's prior use of the phrase, and he did not claim to have coined it. Whitesides & Hoopes, however, have assumed and asserted this and for the record it needs to be corrected.

Here's another:

My work is not treated in the Whitesides & Hoopes article until the final section, titled "Conclusion."[30] Only my 1998 book is cited as support for several dense sentences of assertions. They state that I "promoted" the "ideas" of McKenna and

---

[29] For example, Stray, 2005:239, 288. The original version of my Foreword, completed by March 18, 2004, used the phrase "the 2012 phenomenon." The excised section was originally placed before the second-to-last paragraph, and is here in full: www.Alignment2012.com/the2012phenomenonMarch2004.html. I was unconcerned with cutting it, as I preferred to use the term "2012ology" and called Stray "the first 2012ologist." See www.Alignment2012.com/2012ology.html for more. In the first paragraph of the published version I state "the plethora of writings on it [2012] is a phenomenon in itself" (in Stray, 2005:9).

[30] Sitler described my work as "a central influence on the 2012 phenomenon" (Sitler, 2006:29), so it is odd that Whitesides & Hoopes began what should be a discerning and detailed treatment in the Conclusion to their article, where they treated it superficially.

Argüelles (69). No specific page numbers in my book are offered, and this assertion is demonstrably false. As mentioned, I was the first to publish a detailed critique of Argüelles's systemic errors in 1992. I continued the effort through the 1990s, in 2002, and in my 2009 book *The 2012 Story* (which Hoopes told me in 2009 that he had read), where I also stated my disagreements with McKenna's core notion about 2012 (that a sudden, radical change is to be expected).[31]

And another, after I quote their baseless assertions:

Each of these assertions is contradicted by my published words and efforts, as I show below. The one and only cited source that allegedly supports these assertions is my 1998 book *Maya Cosmogenesis 2012*. Nowhere in that book is there a reference to, or discussion of, a "perennial wisdom tradition" or an interpretive analysis of the Izapan monuments that proceeds "archetypally."[32] They further state that I utilized an "assumption of a pure truth (or insight into the nature of reality) attained prior to cultural dilution, corruption, and textual exegesis." (69) This seems to be a grossly distorted reading of my view that it is best to study the origin place and time of the Long Count (the pre-Classic "Izapan civilization") because it would provide the clearest window into the undiluted original cosmology before historical degenerations inevitably occurred.[33] Their distorted reading of what is a rather commonplace observation about how the passage of time changes the original beliefs of a religious movement or cultural paradigm gives a pejorative slant on my actual approach — which was to study the evidence at the probable origin site of the calendar that gives us the 2012 period-ending date.[34]

None of these assertions by Whitesides and Hoopes are supported by anything that can be found in the cited book source, and I suggest this is a serious breach of academic standards. What they asserted can be obviated with my actual statements taken from that same book. In sharing my actual words below I am not trying to present an argument for my 2012 alignment reconstruction; rather, it is necessary to illustrate that the source cited by Whitesides and Hoopes, allegedly containing support for their contentions, actually provides a completely different picture of my approach and methods, and my consequent evidence-based deductions and interpretations.[35]

---

[31] See, e.g., www.Alignment2012.com/following.html, Jenkins (2009:90-95) for the McKenna critique and Jenkins (2009:104-109) for the Argüelles critique.

[32] Also see my chapter-by-chapter summary of *Maya Cosmogenesis 2012* for a complete lack of evidence for the contentions of Whitesides and Hoopes: www.Alignment2012.com/mc2012summary.html.

[33] I have repeated this rationale many times in my presentations and publications (Jenkins, 2009:151). Michael Coe uses the phrase "Izapan civilization" (Coe 1966).

[34] I found this in Michael Coe's statement: "The priority of Izapa in the very important adoption of the Long Count calendar is quite clear cut" (1988:86). Later scholars concurred (e.g. Rice, 2007).

[35] Relevant quotes and references to the archaeological and astronomical bases of my interpretations are ubiquitous throughout my book, and it's striking that Whitesides and Hoopes selectively ignore them. Five chapters in the Izapa section (Jenkins, 1998:219-

In my Conclusion I sum it up: "…they do not accurately portray my interpretive methodology nor do they cite or address any of the evidence I've brought to bear on my interpretations and reconstruction work (since that is not their concern); instead, they engage in vague citation practices, insinuations of unscientific methods, guilt-by-association constructs, and assertions that are not verifiable and are not supported by the source they cite."

In their response they evaded acknowledging all of the errors I documented and proved in my review. Here is a lengthy excerpt from my counter-rebuttal which includes fair-use quotes from their rebuttal, illustrating their evasive resistance, denial, and clever semantic work-arounds to admitting that any of the errors are actually errors:

As the first example, we can look at my first point of correction, in which I provided the proof for Geoff Stray's earlier use of the "2012 phenomenon" phrase. The published *uses* by Stray were unambiguously *prior* to Sitler's 2006 essay. Recall that the authors credited Sitler with the *first use* of the phrase, as well as with defining the phrase. I also pointed out that there was no definition of the phrase in Sitler's essay. This could have been a simple matter of acknowledging a correction but, instead, Whitesides & Hoopes responded: "We were referring to its use *and* definition." No, they were referring to its *first* use. In my review I quoted their words, that Robert Sitler was "the first to use." They refuse to acknowledge the factual correction on this simple point by saying that Defesche (2007) was "the second" to use it, thus maintaining their demonstrably false position that Sitler was "the first."

Furthermore, they conclude that they "were following and extending this specific scholarly use, not its casual mention" (63). We are apparently suppose to believe that Stray's earlier use (at least four times since 2002 and prior to 2006) was merely a "casual mention" of the phrase and does not constitute a "use of" the term. This fallacious and deceptive way of dismissing a simple factual correction is emblematic of their other responses, and is why I have titled my rebuttal "Deceptive Scholars Refuse to Correct Factual Errors in Their Peer-Reviewed Study."

They also claim that Sitler did define the phrase, and we are suppose to believe that this constitutes a definition: "There is intense and growing speculation concerning the significance of this date among many New Age aficionados and others interested in Mayan culture" (Sitler, 2006: 24). That's a definition? If Sitler's sub-title ("The New Age Appropriation of an Ancient Maya Calendar") is meant to be taken as a definition of the phrase, as they suggest, then we are left with a contradiction in which Sitler himself calls me a "central influence on the 2012 phenomenon" (Sitler 2006:29), yet my work is seen by Sitler to be the most well-researched work on the topic, is recognized as being concerned with reconstructing ancient Maya beliefs and *not concerned* with "appropriating" the

---

298) focus on topography, archaeology, calendrics, site history, and especially astronomical orientation. My analysis of the birth of the Hero Twins (1998:155-166) is based in astronomy, following the work of Tedlock (1985).

Long Count calendar, but with articulating the ancient Maya beliefs associated with it. Thus, Whitesides & Hoopes have misconstrued Sitler's perception of a distinction to be made in the nature of my work, compared to those in the marketplace who have appropriated the Long Count/2012 calendar and have invented various models, doomsday devices, and so on.

In several other examples, Whitesides & Hoopes distort (or ignore) the actual context and *words used* in my corrections *and in their own essay*. Their responses are therefore not to my critiques, but to an imagined distortion or incomplete truncation of my critiques. For example, in their original essay they wrote that I "promoted" the "ideas" of McKenna and Argüelles. I showed proof that I actually critiqued and disagreed with the core 2012 ideas of both McKenna and Argüelles, that I was not *promoting* their ideas, that McKenna himself had made a distinction between our approaches and conclusions, and that my critiques of Argüelles' ideas go back to my 1992 book *Tzolkin*. They responded by addressing NOT their assertion that I "promoted" the ideas of McKenna and Argüelles, but that I merely had an "association" with them and their ideas. See how that works? Obviously, there's a huge difference between actively promoting the ideas of others and simply having some kind of association with those people and their ideas. You could say we were all writers, or males, or had once stood in the same room. That is the kind of chicanery that Hoopes has frequently employed, and it is deceptively employed many times in their responses to my critique of their paper. (See Appendix 3 for full rebuttal.)

In the aftermath of publication, in order to illustrate the events that unfolded, I included in my counter-response one of the emails sent to me from the editor, which had explained to me the terms of my rebuttal. I also included in my support files many of the emails sent to me by Hoopes and Whitesides, including statements made by Whitesides which contradicted what he wrote in his official response. I defend using their emails because it was necessary to illustrate their unethical behavior and unprofessional tactics, both laughable and tragic. Consequently, Hoopes complained to the editor that I was posting private emails. The editor was upset, although the email from him that I used was quite irrelevant and bereft of any kind of questionable content. I sensed that Hoopes had suggested other things to him, completely fabricated, because the editor was quite upset and emailed me that he was going to complain to the association of European journal publishers, or some such agency, and prevent me from ever again publishing anything in European journals. Well, this just confirmed that my job was well done. I had, after all, stormed the Ivory Tower, and it may be that they approved my review — it was recognized and stands as a well-written and documented scholarly piece in a peer-review journal — without knowing I hold no degrees. In my bio, which they requested, I did not in any way state that I held degrees.

So, this particular Experiment proved successful, although Hoopes and Whitesides will forever remain unrepentant and may even rear their ugly heads again in the future. My feeling is that any thinking person who reads their original essay of 2012, my review of 2014 and their rebuttal, cannot in all good conscience help but conclude where the malfeasance lay: squarely in the lap of professional scholars, whose refusal to admit and correct their mistakes was amply demonstrated, and their scientific malpractice got exposed. How do these people walk away with degrees in hand? I guess they put in their time and stumble through what's required of them.

A deeper story no doubt can be speculated upon, of Hoopes and his friend Carlson having discussions as to how they would conspire to mitigate me — Hoopes the author and Carlson green-lighting and defending him. Whitesides being enlisted as a half-concerned junior scholar anxious to get published. Since they are all unwilling to have honest adult conversations with me, or to supply evidence for falsely asserted slanders, or to admit unambiguous errors when exposed, I'm left to wonder if they just followed some half-conscious and uncontrollable urge that I must be mitigated, or if they conspired in a more consciously planned and nefarious way. It doesn't really matter, the story of their collaboration — and I do have some email fragments leaked to me which suggest conspiratorial collaboration — is hereby documented and further detailed in Appendix 3 online; see p. 158.

John Hoopes has abused the peer-review process by using his academic friends to green-light and publish his articles, when they are demonstrably sub-standard and not fit for publishing in an academic peer-review journal. He circularly tries to legitimize his flawed ideas by citing Wikipedia entries that he himself has helped craft. The sources he cites to back up his assertions often fail to provide the evidence for the assertions he made. Worse, his *unprofessional* method of critique is demonstrated in that he often just makes baseless assertions of a damaging nature regarding living authors that never should have gotten through the peer-review process. Attempts to rectify the errors by contacting Hoopes, his editors, and their university press publisher have been met with denial and evasion. All of his peer-reviewed papers on 2012 need to be revised or retracted.

Hoopes's collaboration with Kevin Whitesides suggests an alliance based in a shared agenda (or, at least, a common strategy and goals). The Hoopes-Whitesides component of the Experiment is, like the Aveni example, defining and iconic. I could not have invented a better example of how scholars believe they are above the ethics of their profession, too smart to fail (or get caught), and violate science in favor of shielding their misguided egos from the truth. I do thank the editors at *Zeitschrift für Anomalistik* for seeing the process through, as difficult as it was. And I'd like to thank my friend Kristian Azyndar for his indispensable help and advice.

$$\diamond \quad \diamond \quad \diamond \quad \diamond \quad \diamond$$

John B. Carlson: The Evasive Table Turner

Or, better yet, "he who utters rubbish." My attempts to communicate with John B. Carlson through the years have mainly been a one-way street. Despite this, through our relatively few exchanges and via secondary channels and his own recent writings on the 2012 topic, I've been able to assess his attitude toward 2012 and my work. It now can be shown that his primary ideas about 2012 echo my own interpretations, first articulated many years ago. Paradoxically, however, he dismisses my work as "bogus" and "nonsense" and defends contributors to his journal (namely, John Hoopes) who craft unsupported slanders about my background and ideas. Yes, that's a lot to process, so let's unpack it one piece at a time.

Carlson's *Archaeoastronomy Journal*, of which he is the Editor-in-Chief, is a professional peer-review journal with a board of directors that was published for sixteen years by the University of Texas Press. His education was in radio astronomy and he has done innovative work in understanding ancient Maya astronomy — namely, in how Venus timed "Star Wars" events in the inscriptions. With all these reputable by-lines and professional associations, Carlson must hold to a high standard of non-biased and responsible academic writing, editing, peer-review, and publishing. Well, one would hope so.

My journey with Carlson will take us through 21 years of contact up to the 2015 anthology (which will be discussed in more detail in Chapter 4). There are several levels of culpability at work here. There's Carlson himself as a scholarly writer and the Editor-in-Chief of a peer-reviewed academic journal. Also, the University of Texas Press as the academic publisher of Carlson's journal. Thirdly, the AAUP which legitimizes and oversees the UT Press. This all involves Carlson in one way or another, so let me tell the story from the beginning.

I have a filing cabinet filled with Xerox copies of articles written by scholars. Under "John B. Carlson" I have an editorial he wrote for his journal back in 1993. It applauded the efforts and contributions of independent, self-taught, "amateur" astronomers and archaeoastronomers. I liked that piece. It showed me that there were scholars who might be open to discussing my work. When I sought to present my findings on 2012 astronomy to the community of Mayanists and archaeoastronomers, naturally Carlson's journal was at the top of my list. So in November of 1994 I mailed a query to Carlson's office. After five months there was no response. I may have called and left a message; I can't recall for sure but that's what I'd usually do. In any case, by March of 1995 I was at the Maya Meetings in Austin Texas, and mentioned Carlson's journal to an acquaintance — a scholar who frequently attended the Maya Meetings and was familiar with many of the players. He said off-hand that he knew Carlson and that I should be careful about sharing my original work with scholars, and that I should also make clear when pitching articles that the work is original. I couldn't be sure if his comment was somehow connected, in his experience, with Carlson's prior behavior, or if Carlson was just an unrelated segue for his collegial advice.

When I returned home in late March I sent Carlson a second article pitch, and a third one in June, along with my articles and booklets. I expressly asserted half-way through my cover letter of the third pitch, briefly, that I considered the findings to be my intellectual property. A somewhat stiff way of putting it, but that's what I'd been advised to state, as a protection. And as far as I can tell, and I believe it to this day, the ideas in the pieces I put together to understand 2012 were never articulated before my discoveries. Izapa, the ballcourt alignment, the ballgame's astronomical symbolism, the sun in the Dark Rift, the sacrifice & renewal dialectic between Seven Macaw and One Hunahpu — my "2012 alignment reconstruction" was indeed an unprecedented new discovery in Maya Studies. (Not to mention my reconstruction of Sun-Pleiades-zenith cosmology built into the Pyramid of Kukulcan at Chichen Itza.) Again, no response from Carlson.

In 1997 Carlson was among the short-list of scholars who received my offer to send them a pre-press copy of my forthcoming book, *Maya Cosmogenesis 2012*. I enclosed a self-address-stamped-envelop in my mailing, for the yes or no responses of scholars.[36] Some responded; Carlson did not.

When I gave my presentation that year on my work at the Institute of Maya Studies in Miami (August 1997), Ray Stewart, then president of the IMS, told me he'd asked Carlson for his opinion on my work. Carlson said he was familiar with it and declined to comment. I thus knew he was getting my materials and had enough of an opinion about it to refuse to comment (for whatever reason). Well, his reason became clear the following Spring when I called his office, offering to send him a copy of my new book for review in his journal. He declined, saying he was working on 2012-related research and didn't want to be exposed to other people's ideas. That was a pretty personal response. I didn't know that his journal was basically the John Carlson show; I thought it was perhaps run by a democratic team of people and anyone there might be interested in running a review of my book. Hell, Carlson could just send it on to one of his reviewers and never have to be exposed to the ideas it contained. But no, I must be made to understand that Carlson's eyes would never be cast over its pages and content. For Carlson's work on 2012 was *his* work, unsullied by his exposure to the 2012 work of others.

When my book was published a month later (May 1998) and I had review copies on hand to send out, I sent one to Carlson's office anyway, half-thinking that perhaps he'd change his mind. Never heard from him on that. I felt there was something a little strange going on with Carlson. Maybe the 2012 thing bothered him. When I noticed, in 1999, that his journal would now be published under the auspices of the University of Texas Press, I thought that a more egalitarian change might be afoot in his office, so I sent one final article pitch, which focused on a simple presentation of the Izapa ballcourt's alignment to the December solstice sunrise with my interpretation of the related carved monuments (namely, the throne monument, indicating the ideology of rebirth and world-renewal). No response again. So, I let my efforts with Carlson drop.

Eleven years go by. It's now May of 2010, and I just gave an invited presentation at the 75th conference of the *Society for American Archaeology*, and my well-publicized book *The 2012 Story* had been published the previous October. John B. Carlson announces he is giving a presentation on the "Lord of Maya Creations and 2012" at the Robbins Museum. I read the description of his work and noted that Carlson was articulating what 2012 must have meant to the ancient Maya. I saw he was using many of the same phrases and concepts that I've used in my work. Oddly, he avoided the astronomical component of my work. After his talk one of his audience members sent me an mp3 he made of Carlson's lecture. I was surprised to hear Carlson claim he'd read my 1998 book, was familiar with its

---

[36] I also include my one-page synopsis of my work called "The Astronomy of Baktun 13," which was also sent out with 1997 issues of the Institute of Maya Studies newsletter.

content, but my work was "bogus" and "nonsense" and "by the way, John Major Jenkins and [the others] don't know anything about these things" (such as the Tropical Year-Drift formula). This was the first time I'd heard Carlson comment on my work, after sending him my book, booklets, articles and pitches many times between 1994 and 1999, including a phone call in which I was coolly brushed off. But now I had in hand the unexpected truth. Carlson utters rubbish. I had discussed the Year-Drift Formula in both my 1992 book *Tzolkin* and my 1998 book *Maya Cosmogenesis 2012* — it can be found listed in the Index of both books! Carlson wished to falsely denigrate my knowledge of Maya astronomy in the eyes of his audience, even while the ideas he was presenting were echoing my own. I don't know how else to say it — that's what he was doing. On one hand, he'd told me that he didn't want to be influenced by my ideas, while on the other hand he falsely claims to know what my work is about but finds it to be nonsense.

I contacted Carlson by email and asked for a correction. He turned the tables and claimed I illegally recorded his talk, and I should take the matter up with the directors of the Robbins Museum. Well, I didn't record his talk — someone else did. I took a more diplomatic approach and sent him a cordially-toned laurel wreathe letter, suggesting how we might discuss insights into 2012 that we clearly shared. By this time my friend Geoff Stray — who I had contacted for a second opinion on Carlson's talk — had added a report on his website about Carlson's bizarre statements. I had made Geoff privy to Carlson's ridiculous comment by sending Stray the link to the mp3 file (which the other person had posted as an orphaned blind-link on his server) and asking him for his assessment. For the purpose of his brief review, he took the photo of Carlson off the promotional page for the Robbins Museum talk, and Carlson again claimed thievery (psychological projection, anyone?).

Geoff and I had a little fun behind the scenes and Googled "John B Carlson" to locate other photos, all of various people with the name "John B Carlson". There's one Carlson on a jetski with long hair; there's an African-American lawyer in Dallas; there's a pimply faced high school kid in Oregon.

A few weeks later Geoff and I got a serious letter from the Robbins Museum directors, again accusing Geoff and I of thievery. I responded in all seriousness that they should offer an errata to Carlson's slanderous comments on their website. A series of baffles ensued, in which I had the distinct impression that Carlson was speaking through these people, his friends, by proxy. I wasn't sure why such a cowardly stance was necessary, but the effort broke down because they would not admit any wrongdoing. I offered to send them, and then sent, the mp3 file (in the mail, burned on a CD disk), but they later claimed they didn't receive it. (Ah, the perfect double-bind, to keep me on the gerbil wheel.)

My first letter to Carlson, which he ignored, speaks well enough for my willingness to address his unprofessional behavior in a cordial way, pointing out the shared work and the discussion of ideas about 2012 that we both found to be compelling.

Dear John Carlson,   May 20, 2010

It's been many years since we spoke. With great interest, I looked forward to hearing your report on your God L research, and had a friend record [meaning simply that 'a friend recorded'] your presentation in Middleboro on May 15th. I shared the link with my friend Geoff Stray and he quickly posted what I perceive to be a fair and accurate summary: http://www.diagnosis2012.co.uk/Carlson.html

I want to correct several factually incorrect statements you made, in the hopes that you won't continue spreading misinformation about my work and my knowledge-base of Maya astronomy and calendrics. First, somewhere around the 75-minute mark you defined the tropical year value and the year-drift formula, then added an aside, "...by the way, John Major Jenkins, Pinchbeck, and Argüelles — these guys don't know anything about this stuff." In point of actual fact, both my 1992 book *Tzolkin: Visionary Perspectives and Calendar Studies* and my 1998 book *Maya Cosmogenesis 2012* discuss these topics. You can look up "year drift formula" in the index to either of these books and find the page references. You replied to an audience member's question that you had indeed read my book *Maya Cosmogenesis 2012*. I think not.

Apart from this specific example of you reporting factually incorrect information, I noted several generally cheap shots that are basically meaningless as a rational critique, and only serve a defamatory purpose. I've tried to engage my critics among the "professional Mayanists" on the actual points of arguments and evidence that I've brought to bear on my work, with consistently disappointing results (see http://Update2012.com for the biased assumptions exhibited by Aveni, Freidel, Krupp, Stuart, Malmstrom, and others — all easily exposed). Example after example reveals that you and other critics, exuding an authoritative and informed veneer, are actually extremely underinformed and misinformed about what I'm actually arguing for.

I wish you were willing to think critically about the work your colleagues produce — for example, you lauded Van Stone's recent book but failed to report the several factually incorrect assumptions he harbors. He, like Marc Zender, don't seem to understand the sequence by which Thompson's support for the 285 by 1930 was superceded by his revised 283 by 1950, and then the original 285 was resurrected with flawed arguments by Lounsbury in the 1980s. Van Stone and Zender incorrectly state that Thompson came up with the 283 first and it was later revised to the 285. One consequence of their *faux pas* is the handy rhetorical purpose it can serve, to cast aspersions on the solstice occurrence of the 13-baktun period ending according to the 283 (which is insinuated to be the "old" one, the one that was "corrected"). See how that works? I know that you know the flawed arguments behind Lounsbury's 285, so you might want to help Mark understand the truth; he certainly didn't receive it from me when I clearly and explicitly informed him about the facts of the correlation in our email exchange of early 2008.

You might want to avail yourself of the detailed chronicle of the Long Count's reconstruction, the correlation, the so-called "2012 movement", academic intolerance, and other items I laid out in my recent book *The 2012 Story*. I'd offer to send you a copy but perhaps, as with my offer to send you a copy of *Maya*

Cosmogenesis 2012 twelve years ago, you're doing your own research on 2012 and you don't want to be "influenced by other people's ideas." Well, maybe if you plan on talking about other people's work, you should try to be informed about it. That would be fair.

So, let's have a conversation. My hope is to communicate clearly to you what I've found in 20+ years of investigating Maya astronomy, iconography, archaeoastronomy at Izapa, and 2012-related topics. I find it amusing that all these late-comers in academia are popping up now and can't stand that I've already blazed the trail. For example, one thing I find interesting is that your write-up on how to think of 2012 and God L (as a "First Shaman" figure involved in transformation and worldrenewal) echoes exactly what I've been saying for many years about how the Maya thought of 2012, based on my work at Izapa. It seems likely that the same deity-complex and period-ending ideology that I've identified at Izapa, in the context of the Creation Myth imagery preserved at that pre-Classic site, was preserved into the late-Classic, albeit with some details altered.

I've also been working on the astronomy in the 13 dates on Tortuguero Monument 6, and presented my paper to SAA last month. You'd find this interesting, if you were open to it, as it integrates well with Michael Grofe's insightful work and Sven Gronemeyer's recent observations (in his forthcoming Wayeb Note 34). The 2012 date on Monument 6 is connected in a variety of ways to other dates in the inscriptions and events in Bahlam Ajaw's life. That's where the breakthroughs are happening — and they are affirming both the ideological and astronomical reconstructions I put on the table years ago.

Especially in consideration of your factually incorrect statement regarding my awareness of the year-drift information, and your agreement or disagreement with my correction of it, I'd appreciate a response. And if you'd like a copy of the audio recording of your talk, I will arrange to send you a copy. Best wishes,

John Major Jenkins

Notice my willingness to dialogue. But honest and committed adult conversation must be a two-way street. Carlson doesn't do that; he just turns the tables and evades admitting and taking responsibility for his errors.

The Carlson story was just beginning. It was not my intention but in 2011 he got drawn into the debacle with Hoopes's statements published in Vol. 22 of Carlson's *Archaeoastronomy Journal*. As I detailed in an earlier section, Hoopes asserted I was an astrologer, was therefore engaged in pseudoscience, and that I got my galactic ideas from astrologer Dane Rudhyar. Both were totally false assertions, designed to defame. They were unsupported by any citations or evidence, and this was the crux of my simple request sent to Hoopes — where is your evidence? It's reasonable to request of scholars that they provide evidence for any statements which are defamatory in nature, one of which, in this case, was tantamount to a charge of plagiarism. For a few months in mid-2011 I repeated my requests to Hoopes but he avoided responding. So in September I asked the Journals Manager at the UT Press, Sue Hausman, to help facilitate a resolution. (See the Hoopes section for the Complaint sent to Hausman.) After my official Complaint was sent to her, at her request, she forwarded it to Carlson. His responses were brief and irrational, and he once again tried to turn the tables by asking me about a 1992 book I'd mentioned. In the end, he simply asserted that he found "nothing wrong" with Hoopes's statements.

Carlson had green-lighted Hoopes's contemptuous and defamatory statements, without requiring that proof for them be provided, and then Carlson defended those slanders, seeing nothing wrong with them. The situation was quite amazing and disappointing. Meanwhile, I'd just returned from visiting Tortuguero Monument 6 in Mexico (the "2012" inscription), had just posted my report with my high resolution photos — the best yet available to scholars — only to find that David Stuart's 2012 book was just released (June 2011) in which my Izapa work was, without any discerning analysis or even offering a description of it, dismissed as "mostly nonsense." The year 2011 was filled with these whammies from clueless critics, even while I published breakthrough field research on the Tortuguero 2012 inscription and a new carved monument I documented near Izapa. In addition, in September I released my book *Lord Jaguar's 2012 Inscriptions*.

Meanwhile Carlson published an article in the 2012-themed IAU anthology, in which he further articulated his 2012 interpretations that echoed my own. Again, of course, without any credit given or mentions of my prior work. This evasion of mentioning my work became a mandate in the Carlson playbook, and I have it on record that at least one of his contributors, for the forthcoming Vol 24 "special 2012 edition" of his journal was asked to refrain from citing or mentioning my work. This memo appears to have been drawn from Aveni's anti-JMJ playbook, as we saw earlier, prohibiting his colleagues at the 2008 SAA conference from mentioning my name. We might compare this attitude to the proscription encouraged by County Commissioner Chris Boice, in regard to a recent mass murderer at a school in Roseburg, Oregon: "I challenge you all to never utter his name."[37]

The Volume 24 of *Archaeoastronomy Journal*, edited by Carlson, was released in August of 2012 and contained yet another piece by Carlson, again exploring God L and the Bolon Yokte deity who appears in the Tortuguero Monument 6 "2012" inscription, explicating the "deity sacrifice is necessary for world-renewal in 2012" interpretation that I'd been fine-tuning since the mid-1990s.

By mid-2012 Carlson's inability or unwillingness to communicate with me was clear. However, I noticed that he was now going to comment on the astronomical part of my work, which he'd rarely done before (even thought that should be his primary area of interest as an astronomer). The one prior instance I'm aware of was his mention of it in his 2010 Robbins Museum talk, where he asserted that the occurrence of the galactic alignment in coordination with the end of the 13-Baktun period in 2012 was a "coincidence," absolutely and without a doubt. It's amazing how scholars can be so certain about something they've barely even looked at. His new talk was to be given at the Smithsonian in August, 2012, and it was described in this way:

---

[37] http://wvtf.org/post/never-utter-his-name-official-says-gunman-oregon-shooting#stream/0

103

Speculations about what the ancient sources tell us about 2012 is becoming a global phenomenon in popular culture as the great 5,125-year Maya "Long Count" cycle reaches completion on December 21. How did the ancestral ancient Mesoamerican peoples understand the world in terms of astronomy, conceptions of space and time, calendrical divination, and prophecy? Is it a coincidence that the sun will pass through the plane of the Milky Way near the galactic center around December 21? Did the Maya intentionally create this coincidence? (http://www.alignment2012.com/email-to-Carlson-June14-2012.pdf)

I found this description bizarrely illogical and misleading ... "speculation" in the "popular culture" ... is the alignment a "coincidence"? But he did mention and promise to address my 2012 alignment theory, although of course I wasn't mentioned (at least, not in this program description). I decided to again extend the laurel wreathe to Carlson, and provide him with easy links to my work so that he couldn't get it wrong again, like he did two years earlier in his Robbins Museum presentation of May 2010.

> Dear John Carlson,   June 14, 2012.
> I'm not sure how we would even expect the Maya to "intentionally create this coincidence," but in any case I hope you will accurately summarize what my work is about on this question, and go into some detail about the evidence I've brought to bear on my work. ... [details and links here] ... Please let's try to be honest and accurate. I hope we can work together to move the discourse on reconstructing ancient Maya astronomy forward. Best wishes,
> John Major Jenkins
> (See entire email at: http://www.alignment2012.com/email-to-Carlson-June14-2012.pdf)

I wanted him to know that I was still willing to offer the benefit of the doubt (and I was, in all sincerity). I realized it must have been hard on his ego for me to get there first in the articulation, demonstration, and publication of ideas that he, perhaps, was working out quietly on his own. One can imagine that if Carlson, in 1994 when I first contacted him, wasn't ruled by defensive egoism but by professional ethics, the whole 2012 story might be very different. Carlson, Hoopes, Aveni, and Krupp were all operating from the same short-sighted, guild-protecting, ego-motivated playbook. No one within their guild could be expected to expose such shenanigans, for they'd have their careers ruined. And several of them, who I was close to, did feel the threats and warnings, and muted their own truth-telling accordingly. But I'm not handcuffed by such tactics.

Before the year 2012 ended, there was a panel at the 2012 SAA meeting in May of 2012. Carlson was a contributor, and Aveni moderated it. A short time after it occurred the second 2012 inscription was discovered, this one from La Corona in Guatemala. It promised to blow the lid off the 2012 astronomy question, but the "2012 means nothing" proponents in academia kept the lid on it pretty well. I studied it and wrote three essays on it before July of 2012 was over, showing how a strategy similar to Lord Jaguar's was being employed connecting the birthday astronomy of the two kings to 2012. To this day, more than three years later, those three essay I wrote and published on *The Center for 2012 Studies* website are the only detailed analyses on it that I've seen, apart from David Stuart's initial treatment on his blog, when it was announced.

Carlson's thoughts on the galactic alignment appear to be that he dismisses it as a coincidence. Apparently not even the fact that it falls on a solstice is enough to warrant a closer look. Closed minds, as they say, are a wonderful thing to lose. I'll share more of Carlson's role in the anthology of 2015 in the next chapter. For now, I can only conclude that Carlson has maintained an unprofessional and immature position of refusing to acknowledge how my work anticipated the ideas expressed in his own work. His tactics of communication include evasion, asserting denigrating falsehoods, refusing to acknowledge his misleading statements, green-lighting his friend Hoopes to publish slanderous and unsupported things in his journal, and then defending those unsupported slanders. I guess it's no wonder that in early 2015, the University of Texas Press ceased publishing Carlson's journal, citing incessant and unremedied editorial delays. The publication date of Carlson's volumes have had to be backdated, sometimes by a number of years, causing a deception in the published record — even to the point where authors were citing sources that were published *after* the stated pub date of the authors' own articles! Perhaps Carlson will someday wake up and place the advancement of Maya Studies above the selfish concerns of his own ego.

The 2012 SAA contributors, including Carlson, were tapped to work up their essays for inclusion in an anthology, to be published by none other than the University Press of Colorado, edited by Susan Milbrath and Anne Dowd. It can take years for such a project to be completed. As it turned out, the academic commentary on 2012 wasn't over. The two divergent threads in the discussion had not yet reached polar opposition, for when they did it would signal the emergence of the excluded middle, the synthesis between the thesis-antithesis. We saw how 2012 went, in Coe's 9th edition of his book *The Maya* (released in May of 2015), from Armageddon to nothing. How's that for a polar opposition? In May of 2015 we saw the "corrected" release of Aveni's 2009 book on the 2012 topic, which actually simply enshrined the unacknowledged errors. And also in May of 2015 we saw the release of this incredible anthology, based on the SAA presentations, called *Cosmology, Calendars, and Horizon Based Astronomy in Ancient Mesoamerica*. It was an homage to Anthony Aveni with a Preface by Ed Krupp. Its content and its cover win for it the Ultimate Cognitive Dissonance award, beyond which there can be no more deception. Well, at least I can hope.

Chapter 4. Ultimate Cognitive Dissonance 2015

In this chapter many of the various threads come together in an academic anthology that suddenly appeared in May 2015. I had no inkling it was in the works, even though it was published by the University Press of Colorado and I had been in close contact with the director for several months. As I previously sketched, in May I asked the director for a review copy, but was declined. So I searched around the college libraries in my area and, fortunately, the book was being processed for shelving nearby at the Colorado State University library in Fort Collins. It wasn't yet on the shelf, having just been published, but a helpful librarian found it for me — it had just been processed that day. She handed me the book jacket and said I could keep it as they always remove them and toss them out. This was fortunate because the image on the book jacket, properly understood, is an image of the galactic alignment that has been so reviled and denigrated by Maya scholars and astronomers. And there it was, being used on the cover of the latest scholarly anthology of writings on Maya astronomy, calendars, and cosmology.

I took the book home and read it in two days. Fascinating, great stuff — the kind of nectar that I drank deeply all through my self-directed studies in the 1980s and 90s. The collection was based on the Maya astronomy symposium at the 2012 SAA conference, which was chaired by none other than Anthony Aveni. In fact, the book was basically an homage to Aveni, and I suspect he played a role in getting it published with his friends at the University Press of Colorado. One of Aveni's slightly younger colleagues, Ed Krupp, the director of Griffith Observatory going on forty years, wrote the Preface and highly praised Aveni, specifically noting his 2009 book which took to task the 2012 "End Times Follies." I shared the details on this in the section on Krupp.

The volume was edited by Susan Milbrath and Anne Dowd. Back in 1999 Aveni favorably reviewed Milbrath's first book-length study of Mesoamerican cosmology, called *Star Gods of the Maya* (University of Texas Press). Citing my still-active status as a reviewer for *Colorado Libraries* magazine, I requested and received a review copy. In 2000 I had some email exchanges with Milbrath about her ideas as they compare to my own. Recall that Milbrath was among the few scholars who agreed to receive a pre-press copy of my book, which I sent to her in 1997. I assumed she must have been familiar with it — even if only she read the synopsis on the back cover. But in any case I reiterated some of my findings during our email exchange. Unfortunately she had a few odd convictions about the Milky Way and the Crossroads which partially echoed Aveni's convictions, rooted in the biased reflexes of Western scientism.

Her book came out the year after my book *Maya Cosmogenesis 2012*, and covered some of the same ground, even drawing from many of the same academic sources and arguments that I did in my book. So, it must be said that to some extent we were charting a parallel course in our investigations. However, although Milbrath acknowledges the possibility that the Maya were aware of the precession of the equinoxes, based on the simplistic fact that there are large Distance Numbers in the Dresden Codex, she sees no evidence that they had an accurate estimate of its length or rate of shift. I had already seen her opinion expressed in an earlier article she wrote, and addressed it in my 1998 book:

The November 14 New Fire date of A.D. 1507 was calculated by astronomer E. C. Krupp (1982). Mesoamerican scholar Johanna Broda recognized the dualistic dynamic between the events in May and November (1982b, 1992), but does not draw the potentially controversial conclusion that the Aztecs were really interested in calculating the future convergence of the ever-changing *exact date* of the sun-Pleiades conjunction with the static date of the solar zenith passage. Furthermore, no scholars have recognized that the midnight zenith transit of the Pleiades defines the date of the sun-Pleiades conjunction six months later. Milbrath (1980a) also recognizes the May-November opposition, but, like Broda, only in a general way. (Jenkins 1998:373) [refs: Alignment2012.com/bibbb.htm]

I frequently cited Milbrath's writings in a favorable way, and have always appreciated her insightful work.

The main problem with Milbrath's limited view of Maya precessional knowledge is that she overlooks the level of precision that exists within the ethnographic data. In other words, the account recorded from native informants by Sahagún tells us that the New Fire ceremony was timed by the Pleiades passing through the zenith *at midnight*. It's not clear why this evidence should be disregarded. This midnight criterion is the precise factor that allows us to then recognize how this doctrine indicates the precise conjunction of the sun and the Pleiades, exactly one-half of a year later. From this, and drawing upon other evidence, I derive my reconstruction of what I've called the Zenith Cosmology, which is encoded into the archaeoastronomy of the Pyramid of Kukulcan at Chichen Itza. I devoted several chapters in my book to fleshing out the evidence and argument, reconstructing a previously unrecognized method for tracking precession that points to an alignment in the 21st century. Curiously, the factors involved do not allow for a super-precise fix on the required precessional alignment, so in this sense Milbrath's position that the Maya didn't have the ability for precise alignment calculations is, on this level of my reconstruction, actually moot. But the concepts are there, in the evidence and data.

Milbrath's main issue with my work has been more about her perception of some precise superhuman ability that I supposedly attribute to Maya astronomers, projecting forward to the galactic alignment in era-2012. This came out in our exchange in the pages of the *Institute of Maya Studies Explorer* magazine in 2007, triggered by some comments I made online that were quoted by the magazine's editor, Jim Reed. As a background for Milbrath's role in this chapter, I'll share her article in full, with her permission.[38]

In the November 2007 Issue of the *Institute of Maya Studies Explorer* magazine (Vol. 36, Issue 11), editor Jim Reed wrote a piece called "Understanding 2012"

---

[38] Email of February 3, 2015. She said it was fine, as she had also posted it on her Academia.edu page.

107

which quoted my words, posted several months earlier, responding to a question posed on the Aztlan e-list:

In a post to Aztlan Listserv [April of 2007], a member asks:
In the preceding story [a *USAToday* piece on 2012], note the phrase "On the winter solstice in 2012, the sun will be aligned with the center of the Milky Way for the first time in about 26,000 years." Anyone want to speculated about what this means? I thought that we were always aligned with the center of the Milky Way (as a matter of fact aren't all the stars aligned with the center, since that is the nature of "center").

John Major Jenkins responds:
Allow me to address your question about the solstice sun aligning with the center of our Milky Way galaxy. The alignment referred to in the *USAToday* article refers to an alignment that is caused by the precession of the equinoxes. Precession causes the sun's positions on the equinoxes to shift backward along the ecliptic, moving through the various ecliptic constellations (signs of the zodiac) over many thousands of years. This effect is due to the slow wobble of the earth on its axis. The phenomenon causes the equinox positions - and the solstice positions as well - to periodically come into alignment with the galactic equator - that is, with the bright band of the Milky Way galaxy. It is an astronomical FACT that the position of the December solstice sun will be aligning with the galactic equator in the years around 2012.

Specifically, following calculations by astronomer Jean Meeus (*Mathematical Astronomy Morsels*, 1997:216 [corrected later]), and considering that the sun itself is one-half of a degree wide, we can speak of an alignment "zone", 1980 - 2016 AD. If you would like to read more and see a good illustration of this factual empirical phenomenon, please see my webpage: "What is the Galactic Alignment" (http://www.alignment2012.com/whatisGA.htm).

Secondarily, the alignment of the December solstice sun with the Milky Way's equator happens to occur in that part of the Milky Way that houses the "nuclear bulge" of our galaxy's center. Our Milky Way is saucer shaped, and to naked-eye sky-watchers the Milky Way appears wider between Sagittarius and Scorpio. That "nuclear bulge" is, visually, where the galactic center is located. It is where the December solstice sun is aligning. Thus follows the factually true statement about the sun, on the solstice, aligning with the center of our Milky Way galaxy. It is more precise, however, to speak of the alignment in terms of the galactic equator, as that affords a precise mid-line of the band of the Milky Way with which the solstice-galaxy alignment can be measured – as Jean Meeus did. Thus, the alignment "zone": 1980 - 2016 AD.

My work, since the early 1990s, and published in several books still in print, has marshaled evidence to show that this alignment scenario is the reason why the ancient Maya fixed the end of their 13-baktun cycle to December 21, 2012. For more information, you can peruse my website (http://alignment2012.com), or you can access the Aztlan archives from May-June of 1999, or the compilation of those Aztlan exchanges in a chapter that was excised from my 2002 book

Galactic Alignment: www.alignment2012.com/chapter3.html, or you can simply read my book *Maya Cosmogenesis 2012*, available for free through your institution's Interlibrary Loan service.

This is astronomy, not astrology, not the end of the world. It is a reconstruction of the astronomical knowledge of the early Maya people who devised the Long Count calendar. Susan Milbrath is quoted in the article as stating that it would have been "impossible" for the Maya to have been aware of the precession of the equinoxes. The problem here is that many scholars don't even know what precession is. And, in addition, an awareness of precession is strongly suggested in the work of Gordon Brotherston, Marion Popenoe Hatch, Eva Hunt, and others. "Impossible" is an unnecessarily strong word in this case, as it closes the mind to the possibilities that can be, and have been, explored, elucidated, and documented in my work.

In fairness it should be pointed out that the USAToday piece did not mention my work, *Maya Cosmogenesis 2012* (1998), which is the source for L. Joseph's and D. Pinchbeck's messy but apparently popular books.

Editor's note: *If you would like to read more and see a good illustration of this factual empirical phenomenon, please see Jenkins' webpage: http://Alignment2012.com/whatisGA.htm. Or for more information, you can access the Aztlan Listserv from May-June of 1999, or the compilation of those Aztlan exchanges in a chapter that was excised from his 2002 book* Galactic Alignment, *located at www.alignment2012.com/chapter3.html. Or you can simply read his book* Maya Cosmogenesis 2012, *available for free through your institutions Interlibrary Loan service.*

My comment on Susan Milbrath's statements in the *USA Today* interview prompted her to write a letter of correction which was published in the December 2007 Issue of the IMS *Explorer*. She appears to have read the link offered above, or glanced at it, since in her response she refers to my "alignment zone" interpretation (but calls it "26 years" rather than the correct 36 years). It's hard to understand how Milbrath could even formulate her critique based upon what I actually argue and present in this cited piece. Milbrath's response (in the December 2007 Issue of *IMS Explorer*):

"Just How Precise is Maya Astronomy?"

by Susan Milbrath
Curator of Latin American Art and Archaeology
Florida Museum of Natural History

In Vol. 36, No. 11 of the *IMS newsletter*, John Major Jenkins misrepresents my quote in J. Jeffrey MacDonald's article in USA Today. He says "Susan Milbrath is quoted in the article as stating that would have been 'impossible' for the Maya to have aware of the precession of the equinox." In the 2007 *USA Today* article I say that "astronomers generally agree that it would be impossible for the Maya themselves to have known that [the sun will be aligned with the galactic equator on Dec. 21, 2012.]"

My quote addresses specifically the issue of precision in Maya predictions. As Jenkins knows very well, in *Star Gods of the Maya* (1999:249, 257, 259), I discuss Maya records of long cycles of time that might have been useful in calculating precession of the equinox. The Dresden Codex and certain Classic monuments record pictuns (8,000 x 360 days) and there are also records of much longer cycles of time. In the book, I even refer to a record to a long cycle of 30,000 years involving the Pleiades that may have been an effort to calculate precession of the equinox.

The issue here is that Jenkins implies that the Maya were able to calculate the precession cycle with exact precision. Nowhere do we see a Maya record that accurately records the cycle of precession of the equinox known to us today. This cycle was not known in the West as such until the Renaissance (see James Evans *History and Practice of Ancient Astronomy* 1998). It does seem likely that the Maya recognized that star rise azimuths and heliacal rise/set dates changed through time, but they left no precise records of calculations involving such observations.

Jenkins implies an astonishing level astronomical precision by saying that the Maya were able to predict the exact location of the sun in the background of stars on December 21, 2012. The Maya visualized the Milky Way in quite a different way than we do, seeing it as a road, a river, or a serpent. Without NASA photos they could not have known the true shape of the Milky Way, and it would have been impossible for them to determine the location of the galactic equator. Jenkins tells us that the window for the sun crossing the galactic equator is a scant 26 years, citing Jean Meeus's *Mathematical Astronomy Morsels* (1997:216). This page illustrates only two tables showing the inclination in orbit for two of Jupiter's satellites between 1875 and 2175.

Setting aside the inaccurate page citation, we must ask how the Maya calculated the position of the sun relative to the galactic equator? We can use computer programs and NASA photos to illustrate this effect. Just what mechanism does Jenkins think the Maya used and how did they record this knowledge? Furthermore, tying the 2012 galactic event to the end of the baktun cycle implies that the Maya had precisely calculated precession of the equinox by around AD 300, more than 1000 years before this was achieved in western science. We do know that the Maya purposely set the calendric odometer to "roll over" at end of the baktun cycle on the winter solstice in 2012. This date was predetermined when the first Long Count inscriptions were recorded in the third century AD in the Maya lowlands (even earlier in the areas of Veracruz and Chiapas).

The end of the baktun on the winter solstice is not a coincidence, and this mathematical feat is certainly a sign of a sophisticated link between Maya astronomy and mathematics. The Maya must have set the baktun "end" at the same time they back-calculated a starting point for the baktun around 3000 BC. We can admire the Maya for their highly developed astronomy and mathematics, but we should not attribute to them impossible feats and thereby diminish their true accomplishments.

[Note: The editor added one small image of the sun at the Crossroads/Dark Rift (the 2012 alignment), with no caption.]

My response to Milbrath was then published in the February 2008 issue of the IMS *Explorer* (IMS Vol. 37, Issue 2). Section headings were added by the IMS magazine editor, Jim Reed.

December 21, 2012: Some Rational Deductions

Editor's note. In November 2007 the IMS published an article about the 2012 phenomena in which John Major Jenkins commented on statements made by Maya scholar Susan Milbrath in a 2012 article previously published in USA Today. *Susan responded to John's comments in an article we published in our December 2007 issue. Now, read on as John continues the discourse.*

I appreciate Susan Milbrath's thoughts on Maya astronomy, and hope our exchange will inform and inspire the investigations of others. The citation I provided off the top of my head for Jean Meeus's calculation of the solstice alignment with the galactic equator was indeed incorrect. It's not on page 216 of *Mathematical Astronomy Morsels* (1997) as stated, but pages 301-2 of that book.

Milbrath clarifies her intent in the quotation that I had misread from the somewhat unclear phrasing of the USA Today journalist. Instead of stating that it would have been impossible for the ancient Maya to know about precession, she means it would have been impossible for them to have been aware of the galactic alignment of era-2012 with the level of accuracy she believes I require. Here, however, she mistakenly imputes that I require a level of precision that I do not. The confusion seems to arise from my use of precise astronomical terminology to define what the galactic alignment is. I define the galactic alignment as *the alignment of the December solstice sun with the galactic equator.*

An Alignment Zone?

Based on Meeus's precise calculation for the galactic alignment occurring in 1998, and the fact that the sun is roughly **one-half** of a degree wide, I pointed out that it is reasonable to think of this alignment as a "zone" stretching from 1980 to 2016 (36 years equals **one-half** of a degree of precessional motion). I then observe the fact that the 13-Baktun cycle ending date falls within this range. In the early 1990s, I took this more generalized situation of "being in the zone" not as a definitive proof of my alignment thesis, but as an indicator of *possible* intent on the part of the Long Count's creators, and a suggestion that it might be worthwhile to rationally investigate the topic further.

I think that my use of a scientifically accurate definition of the galactic alignment has been conflated with what ancient Maya naked eye star gazers would have been, and could have been, looking at. To use the term "galactic equator" in a scientifically precise definition of the galactic alignment does not mean that the ancient Maya star gazers utilized the exact same scientific concept that modern astronomers do. It does not logically follow. Thus, Anthony Aveni's challenge (as reported by Ben Anastas in the New York Times article of July 1, 2007: http://www.alignment2012.com/NYTimes.html) to demonstrate a Maya awareness of the galactic equator simply misses the point. I neither assume this nor state this in my investigation of ancient Maya astronomy.

What is significant, however, is that certain astronomical features that are compelling *to the naked eye* are involved in the galactic alignment (the Milky Way, the dark rift in the Milky Way, the cross formed by the Milky Way and the ecliptic, and the sun) and are very important players in the Maya Creation Myth. It should also be emphasized, as I've frequently stated in my published work since the mid-1990s, that the Milky Way itself, and more narrowly the dark rift in the Milky Way, would have served the ancient naked-eye astronomers as the target for the galactic alignment rather than the abstract dotted line known to astronomers as the galactic equator.

A Conceptual Awareness

Another assumption that frequently occurs should be clarified. Instead of honing in on the ancient Maya's observational and calculational methodology and assuming a high level of precision, my approach to demonstrating 2012 as being intentionally placed proceeds along different lines. (I'm sure that progressive scholars will one day identify precessional interval mechanisms in the dated hieroglyphic corpus, as well as models of how precessional ideology relates to kingship and other much-discussed features of Maya culture, but I have approached the problem in a different way).

My methodology endeavors to show a meaningful presence, within core Maya institutions, of the astronomical features involved in the galactic alignment. This coordination involves several disciplines—it is an interdisciplinary synthesis—the integrative continuity and complexity of which mitigates the possibility that my observations are all just wishful thinking.

So, it is a false assumption that calculational methodology must be provided in order to prove intention. My methodology documents the secondary effects that are predicated upon and require an ancient awareness of the galactic alignment. The deduction is similar to deducing that Paleolithic humans knew how to have sex, because the secondary effect of that knowledge—progeny—is observable. We don't have any direct evidence for their knowledge or their "methodology" but the secondary evidence that they knew how to do it is undeniable.

Similarly, the astronomy of the galactic alignment is embedded within the Maya Creation Myth, on the monuments of Izapa, and in other Maya traditions such as the ballgame. A conceptual awareness of the alignment is clear in the iconography and the symbolic representations of astronomy within Maya mythology. This brief response to Milbrath is not the place to go further into the details of my reconstruction; the point is that my investigative methodology does not hinge upon the high level of precision that some criticize as being "impossible."

In my books I've even stated that 100 years within range of the precise galactic alignment calculated by Jean Meeus would still be compelling enough to justify further investigation.

It Isn't a Coincidence

I've addressed and clarified these issues on the Aztlan listserv, the University of Texas Mesoamerican e-list page and elsewhere, including a brief piece I wrote during recent exchanges with professor John Hoopes (see link below). This online article revisits the suggestion that it is unlikely for the solstice placement of the end date to have been a coincidence, an idea supported by Milbrath in her rebuttal. Munro Edmonson pointed this out in his 1988 book *The Book of the Year*. I have explored and restated the implications of this idea and the interested reader can assess a rational analysis of the situation here: http://Alignment2012.com/rationalapproachto2012.html.

—end

That was the last I heard from Milbrath until our communication in 2015, regarding the anthology she edited with Anne Dowd for the University Press of Colorado. The link at the end of my response to Milbrath shares the full treatment that I submitted, but which could not be fully reproduced due to space limitations. But it serves as my full thoughts on the issues raised. One of the ideas I explored, which I first noted in my 1994 book *Mayan Sacred Science*, is the idea that 3114 BC and 2012 AD are like-in-kind Creation Days, bookends of the Maya World Age doctrine. As such, the deities and Creation events involved in one could be expect to reappear in the other. But it wasn't quite so simple. I noted that both dates involved the sun being positioned within a "cosmic center" — the zenith at the earlier date and the galactic center in 2012. That essay has been on my website since early 2008. It was relevant to the Milbrath discussion and was based on an essay I had just written at the request of John Hoopes, to explain my work.[39]

---

[39] Ironically, Hoopes read it and called it the best treatment he'd seen from me, but then proceeded, exactly at this time, to busily craft his "Mayanism" entry of Wikipedia, where the idea of World Ages was indicted as a non-Maya New Age intrusion, injected by either Franciscan translators or nefarious outsiders into the 2012 discussion and part of what Hoopes called an invented 2012 mythology. This was Hoopes's response to losing an argument with me about evidence for World Ages in Mesoamerican traditions, in an exchange we had in late 2007. Rather than acknowledge the evidence, instead he crafted a polemically twisted counter-position on Wikipedia, exploiting the fact that the World Age doctrine is part of "New Age" thinking. See how the profane intellect, the mind of debunkery, works?

I didn't hear anything more from Milbrath after our exchange in the pages of the IMS *Explorer*. I reached out in early 2015 to ask permission to use her mini-article for my writing project, and she agreed. I asked her about an unclear comment made by Victoria Bricker in one of her books. The exchange was brief. Then, a few months later, I read the new anthology she co-edited. As mentioned, it was exciting to see such focus on elements of Maya astronomy that I hold dear. The year 2012 was mentioned several times by different contributors. This was not surprising, as the SAA conference where these papers were presented took place in mid-2012, and the impending date was on everyone's minds. Most notably, one of the editors, Anne Dowd, mentioned it in a causal way as, of course, being about renewal. Ed Krupp dismissed it in his preface, and John B Carlson mentioned it in his chapter. Carlson's piece explored the so-called Flood scene from the Dresden Codex, concluding it wasn't intended to be a 2012 apocalypse scene, as some in the pop marketplace have asserted. He also reiterated "his" interpretation of period-endings like 2012 being about deity sacrifice and renewal. And, adding a third factor, about human "participation" in the process. This third factor also echoes how I've interpreted the ballgame as a metaphor for the 2012 alignment — that we, like the ballplayers, are indispensable participants in the process of renewal, because we choose to sacrifice "Seven Macaw" (i.e., that which is corrupt, deceptive, and illusion).

I was also struck by the chapter from Clemency Coggins, who I'd corresponded with about my work in 1996. While Carlson reiterated the ideological part of my reconstruction, Coggins partially echoed some of the astronomical ideas on precession that I reconstructed at Izapa. She certainly didn't broach the subject of the galactic alignment, but she concluded, as I had, that it was likely that the pre-Classic people knew about precession. She also described a likely process that I had first enunciated in my 1998 book — that as people migrated southward into the Americas, the focus on the Polar Region as the throne of a preferred cosmic center deity became less valid, as that region is low on the horizon within the Tropics. In my work I suggested that this put pressure on the skywatchers to locate a new cosmic center, so the process involves a shift or evolution in cosmological focus. She also points out an up-down motion on Izapa monuments, related to the rise and fall of stars over the North Celestial Pole. I also suggested this, and also noted that the Big Dipper rises over Tacana volcano to the north — there are diagrams in my 1998 book illustrating the process. Coggins has a diagram of the northern horizon with the Big Dipper stars rising, but she neglected to depict the topography of Tacana volcano — which is really the critical piece of data because of the shallow cleft on the eastern flank of the volcano, serving as an "emergence place" for the Big Dipper.

So, collaboration among researchers can result in a larger consensus, and I communicated these things to Coggins in a few emails in May 2015. My first and second, with her reply:

Dear Clemency Coggins,   May 22, 2015

I recently read with great interest your chapter "The North Celestial Pole in Ancient Mesoamerica" and was fascinated to find you drawing interpretations similar to ones I offered in my 1998 book *Maya Cosmogenesis 2012*. You may recall that we communicated briefly in 1996 (maybe other times, too?). It is certainly a long overdue confirmation that my interpretation of the geographical migration of the polar region as a cosmic center deity, from the far north into Mesoamerica, is not as wild as critics have characterized it as being, and will in fact be readily apparent to any scholar who takes the time to study pre-Classic tropical astronomy, sites, mythology, and the topography around Izapa.

I was hoping to discuss with you other elements of Izapa cosmology and pre-Classic precessional knowledge, that I believe are congruent extensions of the ideas you offered in your article. In particular, they involve a more complete reconstruction of ancient cosmology and astronomy at Izapa, by incorporating the iconographic and archaeoastronomical evidence from Groups A, B, and, importantly, the Group F ballcourt.

I have not seen incorporated into any discussions of Izapa cosmology (including Guernsey) the fact of the ballcourt's alignment to the December solstice sunrise horizon (Jenkins 1996, 1998). In 2009 I became aware of Aveni & Hartung's confirming publication of the ballcourt alignment, in their essay published in 2000. (I wonder why Aveni didn't tell me about it earlier, despite our communications in 1996-1998 and 2007-2008?) In any case, my independent discovery, calculation, and first publication of it was in my 1996 monograph *Izapa Cosmos*, then in *Maya Cosmogenesis 2012* (1998). Oddly, in Aveni's 2009 book (*2012: The End of Time*) he reports the Izapa ballcourt alignment <u>48 degrees</u> in error.

In my 1998 book I cited eight essays by you (see bibliography here: http://alignment2012.com/bibbb.htm), essays that I found very insightful.

Another important piece of the Izapa puzzle, which is critical for understanding my work and that I haven't seen broached elsewhere, is the fact that the Big Dipper rose over Tacana volcano at sundown on December solstices during Izapa's heyday (Jenkins 1998: Diagram 138, p. 255). This fact would have been apparent if, in your Figure 5.16 (p. 128), you had portrayed Tacana volcano in your horizon astronomy map. I published this and many other things in my 1998 book, which contains five chapters devoted to analyzing Izapa's calendrics, archaeoastronomy, topography, regional ideas, archaeology, and iconography. I've been to the site over a dozen times since 1990.

Would you like to discuss things further? Best wishes, and thank you for your work,

John Major Jenkins

Coggins responded:

May 22, 2015

I am very sorry John if you feel I have stolen some of your ideas. I do not think so. However our correspondence was long ago and I suspect I agreed with many things you said. Unfortunately I have never seen your book *Izapa Cosmos,* although I did see *Maya Cosmogenesis* around 2000 and apologize for not going back to it.

All the initial work for this paper was done long ago for an SAA panel in San Diego. Any up-dating I did involved the most recent published work I could find. I have never thought much about the north group because I wanted to keep my points as simple as possible. I imagine those monuments do have astronomical significance. I was lucky to get that single Izapa sky figure into the article. They kept cutting down. Had I an artist, or talent, it would have been good to include Tacana.

In any case my interest is, primarily, the evolution and development of Maya iconography and symbolism accross the millennia. There is more at Izapa than I had room for certainly. Perhaps I will get to it - and be sure to credit you.

Clemency

Clemency Coggins Professor Emerita
Departments of Archaeology and of the History of Art and Architecture Boston University. Research Associate Peabody Museum, Harvard University

My responses to Coggins:

May 22, 2015

Hi — oh, I wasn't assuming that at all. I was rather delighted that your research had brought you into similar interpretations, based on your reasoned assessment of the evidence. I've tried to share the work with other researchers, often with inconclusive results. There's a perception that Group F may be a much later occupation, but this comes largely from Garth Norman's somewhat forced model. It's true that there was Classic and post-Classic activity in the ballcourt group, but the closest C-14 dates in the attached Mound 125a are pre-Classic.

For some reason Izapa is often overlooked or incompletely interpreted. I'm not sure why; perhaps a perception that it's BYU's baby. I'm grateful you have studied and written about it. The Izapa ballcourt alignment is quite frequently mis-portrayed in the literature, due to the unclear maps in the BYU studies. The monuments arranged around and in the ballcourt are congruent with the solar rebirth symbolism of the ballgame, and I briefly summarized this work in my 2010 SAA presentation paper, which otherwise was about Tortuguero Monument 6. That paper was posted on Barnhart's *Maya Exploration Center*, and on my website: http://www.thecenterfor2012studies.com/Astronomy-in-TRT-SAA.pdf. The paper was put through the ringer in a public debate that Barnhart sponsored (http://www.thecenterfor2012studies.com/MEC-Facebook-Discussion-2010-ON-Jenkins-SAA-TRT-Astronomy.pdf) and developed into a 9000-word essay to be published with an archaeoastronomy anthology edited by Benfer & Adkins. But it remains unpublished, though an abridged and rewritten version was published in the Clavis Arts Journal in 2014 (http://thecenterfor2012studies.com/Clavis3-description.pdf).

Also, important work by Michael Grofe is being done on Maya awareness of the precession of the equinoxes. I hope we can stay in contact about developing discoveries. Best wishes,

John

There was no response after this. It was nice to receive an acknowledgement of our prior exchange in 1996, and that she would hold my work in mind in the future. In some ways this minimal effort is all I've ever expected from any scholar. But as we've seen, most scholars have been unwilling to credit and acknowledge even when my work has been shared with them, and often go further to denigrate and discredit through malicious means even while their interpretations come to reflect my own.

Coggins' reflex that I was suggesting she had "stolen" my work, which was not my assumption, is echoed in my subsequent exchange with Susan Milbrath — even though I explicitly said I wasn't assuming that Carlson or Coggins were plagiarizing my work. It's worth reproducing my cover letter in full, with the ensuing emails. My lengthy "review" of the anthology can itself can be found in Appendix 3 online (see p. 158). I should emphasize that everything started to converge at this point, since in addition to Milbrath the ensuing events brought in Krupp, Carlson, and Hoopes, while my second phase of emails with Pratt was wrapping up, about Aveni's "revised" eBook. And, to top it off, my first exchange with Morrison happened in June. Like I've said, this was like playing five chess games at the same time.

Dear Susan Milbrath and Anne Dowd,   June 3, 2015

I am enclosing an adapted and personalized version of my forthcoming review-essay of *Cosmology, Calendars, and Horizon-Based Astronomy in Ancient Mesoamerica*. Overall, it is a welcome contribution to the field, and congratulations! But there are a few problems. I've learned over the years to not expect too much in the way of dialogue with scholars and critics, even if they have directly and publicly commented misleadingly on my work. However, I really hope you will seriously consider my concerns articulated in my enclosed essay.

It should be of interest and a concern to you because, in essence, you should come away with the awareness that "oh my, in the anthology of articles that we supervised and edited, two of the contributors have echoed, and presented as their own proposals, ideas that were actually argued, articulated, and published long ago by another researcher." This circumstance is compounded by the fact that I had communicated with and sent my work to both of the relevant contributors (Carlson and Coggins) in the 1990s. I will offer that, in a recent cordial communiqué with Coggins about her article, she acknowledged that she apparently overlooked how some of her ideas reiterated my own, and offered to acknowledge my work in a future publication. Carlson's relation to my work is a more difficult matter, which my essay explains.

There are also difficult implications, that you should be made aware of, with the denigrating comment by Krupp, regarding the 2012 "End Times Follies," as well as the frequent and favorable citations to Aveni's 2009 book *2012: The End of Time.*

I offer an informed and concerned critique of your book — the enclosed customized excerpt contains only a partial review that focuses on selected aspects of it. I hope that one or both of you will offer a considered response. I can boil this down to shorter treatments for publishing in various places, and I'd like to be able to report that the editors have offered a considered response. Generally, I'm happy that these ideas are getting out there; we just need to get the proper credit and sequence of discovery correct, which I'm sure you will agree is important for the published record. Best wishes,

John Major Jenkins

I pasted my lengthy review into the body of the email and also attached it as an MS Word doc. It's all in Appendix 3 (p. 158), but I'll reproduce below the final section:

Final Note: The cognitive dissonance in the University Press of Colorado's behavior is underscored in their two releases of May 2015: 1) Aveni's allegedly corrected book of 2009 and 2) the new anthology, *Cosmology, Calendars, and Horizon-Based Astronomy in Ancient Mesoamerica* (edited by Anne Dowd and Susan Milbrath). The former refused to register timely corrections while maintaining false and damaging assertions about my religious persuasion and my scholarly work; the latter both cites Aveni's 2009 book as a viable and reliable source of critique on the 2012 "follies" (Krupp) while other contributors echo my core interpretations of 2012, in both the astronomical and ideological aspects (Carlson, Coggins). And, strikingly, the book's very cover image depicts the same kind of solar alignment with the Milky Way, highlighting the Dark Rift in the Milky Way aligned with the sun, that is the centerpiece of my much denigrated and dismissed reconstruction work, published since the mid-1990s.

Dowd never responded. Milbrath responded curtly:

Sorry John that you are offended by not being cited. Scholarlship is always evolving and sometimes people even come up with the same ideas. I am not interested in getting into whether there is any truth to your claim that these scholars are stealing your ideas. Because you send scholars a self-published book does not mean that they ever read it. In fact, many people may just toss unsolicited materials or block the sender, in the case of email.

Susan Milbrath, Ph.D.
Curator of Latin American Art and Archaeology
Florida Museum of Natural History
Dickinson Hall, Museum Road
University of Florida, Box 117800

Gainesville, FL 32611-7800

Her misreading of my actual words was rather surprising, and I sought to clarify:

Susan,   June 4, 2015
You wrote: "I am not interested in getting into whether there is any truth to your claim that these scholars are stealing your ideas."

I did not claim that these scholars were stealing my ideas. Did you read my email? I didn't conclude or assert that Carlson or Coggins were "stealing" my ideas. I will be happy to re-quote what I actually stated in my email, if you need me to. What can be said is that both of them, having by their own admission been demonstrably aware of my work for many years prior to SAA 2012, forgot or, in Carlson's case, intentionally neglected to cite or mention my earlier interpretations which anticipated his own. That is demonstrable. Will the published record correct itself? Well, only if those editors, authors, and publishers who are responsible for upholding academic standards are willing to be fair, principled, and reasonable. So far the track record is pretty appalling.

One issue to raise with your anthology is that, given that your contributors are, by whatever route of their own research, echoing my long-ago published interpretation of 2012 and pre-Classic Izapa, how do you and your publisher reconcile the simultaneous denigration of my work, courtesy of Krupp and citations to Aveni's demonstrably flawed book of 2009? I provided the details in my email. Do you see the contradiction? That is one thing I'm trying to point out here, so the publishing record does not continue to write me out of the narrative while my work gets revived and adopted into the consensus through others.

It was proven that Carlson received my article submissions and proposals in the mid-to-late 1990s, because he had xeroxed a piece I had sent him and he sent that later to John Hoopes. Also, I talked to Carlson on the phone in 1998.

The spiral-bound book (and essays) I sent to you and other scholars in 1997-98 was the prototype that became, with a few alterations, the published book *Maya Cosmogenesis 2012* (1998). It was a trade book, exactly like 4 of the 5 books on 2012 published by scholars that are frequently cited. It was not, like Van Stone's 2012 book, a "self-published" book. While you might not have read it, or even opened it, our email conversations in 2000 show that you were aware of my work. This kind of direct communication with other scholars also occurred with, for example, Carlson, Coggins, Krupp, and Aveni.

I'm sorry if this is an uncomfortable conversation. I feel I'm being diplomatic and my comments and questions are reasonable. I wouldn't bring them to your attention, or publish them in a review-essay, if they weren't well-grounded in facts and evidence.

So, while it's understandable you wouldn't want to address my supposed "claim" that my ideas were stolen (which, again, is NOT what I stated), perhaps you can suggest a proactive process by which the ideas central to my work that are expressed in your anthology can be discussed and debated, as well as some way that my other concerns and observations can be addressed by responsible scholars in a professional way? For, the fact remains that your anthology simultaneously denigrates the 2012 "End Times Follies", for which (according to Krupp) I am supposedly the main choreographer, while my core ideas about 2012 and Izapa cosmology are being echoed in two other articles in the anthology (Carlson and Coggins).

Can there be any acknowledgment that you are hearing and understand my concerns and critique, or is there just going to be a flat-out rejection that they are reasonable and valid? Sincerely,

John Major Jenkins

Milbrath responded:

June 4
Sorry John, that is just what you do say below—you accuse John Carlson of stealing your ideas: "in Carlson's case, [he] intentionally neglected to cite or mention my earlier interpretations which anticipated his own." Then you say: "It was proven that Carlson received my article submissions and proposals in the mid-to-late 1990s, because he had xeroxed a piece I had sent him and he sent that later to John Hoopes. Also, I talked to Carlson on the phone in 1998."

Then you accuse us as editors of not being fair and reasonable, suggesting we should have had John acknowledge a debt to you: "Will the published record correct itself? Well, only if those editors, authors, and publishers who are responsible for upholding academic standards are willing to be fair, principled, and reasonable." In other words Anne and I are accused of not upholding academic standards. Then you go on to say: "how do you and your publisher reconcile the simultaneous denigration of my work, courtesy of Krupp and citations to Aveni's demonstrably flawed book of 2009?" I am copying this to John Carlson, as I do think he should know what you are saying about him. I won't copy this to Tony and Ed, as I am sure they long ago stopped communicating with you, and now I see why. Signing off for good.

Susan Milbrath, Ph.D.

"Signing off for good" was often how scholars sent the signal that they could not handle the facts being presented to them, and were not willing to logically process clearly presented information. It was strategically better for them to be perceived as "misunderstanding" than to acknowledge the facts.

Dear Susan,   June 4
You have fully misunderstood my statements and have jumped to a wrong conclusion … [Here I placed the email I sent to Milbrath and John Carlson, which is in full in **Item 5e** in Appendix 3 online; see link on p. 158. There was no response from either Milbrath or Carlson to this.]

I found Milbrath's irrational insistence on misreading my actual words, and her unwillingness to acknowledge what I'd actually stated from the very beginning of our correspondence, to be perplexing and unprofessional. I tried to help her perceive what I'd actually stated by restating what I actually stated and which she refused to acknowledge:

[To Susan Milbrath]   June 5
Briefly, it may be helpful if I do re-quote a passage from my review-essay, sent in my first email to you:

> "It's not clear if Carlson directly plagiarized my idea, **or if he was suspecting the same interpretation about 2012 that I had explicated**, documented, argued, and published much earlier than he did..."

If you could get past the erroneous fixation that I've unambiguously claimed that "stealing" of my ideas occurred, we could probably have a productive conversation.

>   John Major Jenkins

Due to the unwillingness of a scholar to engage in dialogue, my effort had to be dropped. However, I had communicated with Ed Krupp and felt I should share his current comments to me about my work with Dowd and Milbrath, as it confirmed my suspicions:

Dear Susan and Anne,   June 13, 2015
I recently communicated with Ed Krupp, and confirmed that he does indeed see me as being part of the 2012 Maya calendar "End Times Follies" that he alluded to in his Preface to your anthology. He wrote in his email to me (June 9, 2015):

> There is ample documentation to associate your first book and related activities with all of what I actually called the "2012 Maya Calendar End Times Follies." …If I were to write another piece specifically on the 2012 theme, it would be folly to omit you. [—Ed Krupp]

His 2014 article in *Handbook of Archaeoastronomy and Ethnoastronomy*, as well as an updated version of his 2009 *Sky & Telescope* article (published in *iQ Magazine*, December 2012), also make it clear that he believes my work was a primary foundation and player in this "End Times Follies." While he qualified that this doesn't necessarily mean I advocated doomsday, his Beckman Center talk of November 2009 certainly gives this impression.

He didn't explicitly mention me in his Preface, apparently due to brevity considerations and his focus on Aveni's work. But, my point is that his allusion to the End Times Follies is simply a short-hand nod to the allegedly dubious work of myself and others. The issue here is that he's constructed a category that disallows my work on 2012 to be relevant to "real" academic work. No contribution to our understanding of 2012 can be found there. However, as I've tried to convey to you, the main point of me contacting you was to underscore a contradiction that exists in your anthology. That being the fact that two of your other contributors echo my long-ago published ideas on 2012, Izapa cosmology, and pre-Classic precessional cosmology (Carlson and Coggins). These are facts. You either disagree with the facts or won't acknowledge this contradiction, both of which I find unacceptable.

Importantly, I'd like to get your assurance that you understand that I was not accusing Carlson or Coggins of plagiarism, as you insisted I did. My position was explicitly clear in passages that I wrote in my email to you, such as:

"It's not clear if Carlson directly plagiarized my idea, or if he was suspecting the same interpretation about 2012 that I had explicated, documented, argued, and published much earlier than he did..."

"...in a recent cordial communique with Coggins about her article, she acknowledged that she apparently overlooked how some of her ideas reiterated my own, and offered to acknowledge my work in a future publication."

I also sent Carlson the full review-essay I had sent you, so he can be fully informed of my observations (I wasn't sure if you had forwarded the entire thing to him when you said you were informing him of my supposed "accusations"). I also queried him regarding the similar ideology we both found, that relates to 2012. As I already mentioned, I feel this could be fleshed out as pre-Classic and Classic Period inflections of the same underlying period-ending doctrine (deity sacrifice and world-renewal). But no response, as usual, from Carlson. His huge ego apparently prevents him from acknowledging my work so that Maya Studies can progress. Instead, he and his buddy Hoopes continue to craft mitigations and work-arounds so that my prior pioneering work gets buried.

I was hoping that other scholars would not abet or participate in this deception, and that at least a few scholars might be able to be honest and fair, and could respond to facts and evidence as to the sequence of discovery and publication of certain ideas about 2012. One need look no further than my own SAA presentation, in 2010, which preceded by over two years the SAA 2012 presentations that comprise your anthology. But if you wanted to look further, we could trace my interpretations on 2012 back to *Maya Cosmogenesis 2012* (1998) and my 1997 *Institute of Maya Studies* presentation ("The Astronomy of Baktun 13"), which is now transcribed.

So, the un-answered question remains, as I wrote in my earlier email:

"One issue to raise with your anthology is that, given that your contributors are, by whatever route of their own research, echoing my long-ago published interpretation of 2012 and pre-Classic Izapa, how do you and your publisher reconcile the simultaneous denigration of my work, courtesy of Krupp and citations to Aveni's demonstrably flawed book of 2009?"

The bolded part emphasizes, again, that I wasn't assuming plagiarism. Sincerely, John Major Jenkins

No response from Milbrath. Krupp, as previously discussed, stumbled away from responding to my exposé of his error-riddled articles and presentations, saying he needed to "take a time out." There was no response from Carlson, of course, but as I discussed in a previous section, I noticed that Carlson and Hoopes were, precisely at this time, accessing one of my essays on my Academia.edu page — the one called "Mayanism: An Ideological Prison Created by John Hoopes."

Why does the new anthology of 2015 represent the Ultimate Cognitive Dissonance? Well, you have two scholars in two different chapters echoing my unique interpretations of 2012 and related precessional ideas. So, both the ideological and astronomical aspects of my pioneering work from the 1990s have been reiterated without, of course, any credit given or mention of my work. Meanwhile — here's the cognitive dissonance part — Krupp indicted me as a primary contributor to his 2012 "End Time Follies" and lauds Aveni's book for debunking 2012. It's true that in this instance Krupp's allusion to me was indirect, due to reasons of concision, but I confirmed that he would normally include me in a fuller critique (as he did in his 2014 article) — and he said it "would be folly" not to do so. The anthology thus reveals the state of academia in regard to my work. It unconsciously recognizes the merit of my work because several key elements of it have been reiterated by other scholars, yet Krupp's allusion to the Follies as well as the many kudos given to Aveni's 2009 book, which is deeply flawed but often cited as a valid debunking of my work, consciously casts my work aside. The 2015 book contains a simultaneous dismissal and reiteration of my work, one conscious and one unconscious — that's cognitive dissonance.

The "ultimate" part comes from the portrayal of the galactic alignment on the book's cover — the very thing that scholars almost universally chide, ridicule, and dismiss. For one cannot imagine a higher or more lofty paradox. Perhaps you shouldn't judge a book by its cover, but then again a picture's worth a thousand words. It's an image of the very centerpiece of my reconstruction.[40] Absurd? Hilarious? Pathetic? Disgusting? Acceptable? Discerning readers can judge for themselves.

I look back over these events, laid out clearly in this narrative, and am amazed at the dearth of rational intelligence, honesty, and integrity within Maya Studies — at least, in relation to the issue of my work and the 2012 topic. This was the perfect arena to test whether or not science would function and do its job. New ideas need to be rationally and honestly assessed, with unbiased critical discernment operational. But time and time again, instead we see irrational behavior, biased debunkery, malicious *ad hominem* jabs, and dishonesty. What's going on here? Well, we've been storming the Ivory Tower, and now can see that it is nothing

---

[40] It's also virtually the same image used by Grofe (2011) to illustrate the birth of God I on the Tablet within Palenque's Temple of the Cross. See my explanation in my SAA presentation of April 2010 (Jenkins 2010), as well as in my review of Stuart's book on www.Update2012.com (in the David Stuart section).

more than a house of cards, a flimsy structure of fragile egos and fake posturing, baseless assertions, bad scholarship, and opinionated denunciations.

❁ ❁ ❁ ❁ ❁

Chapter 5. Checkmate: When the Ivory Tower becomes a House of Cards

Actually, rather than *storming* the Ivory Tower, I feel like I've been patiently knocking on the door, ringing the bell, peering cautiously through dingy windows, looking for signs of life. "Hello, anyone home? It's progress calling!" We seem to be at the same stale impasse that Maya Studies experienced in the early 1970s, during the final years of the life of J. Eric S. Thompson. He prevented progress in deciphering the Maya script because he resisted the contributions of two Russians (Tatiana Proskouriakoff and Yuri Knorosov). That bigotry is almost exactly mirrored in the bigotry evident in Aveni's 2009 book, which crafts a perception of me being a member of Gnosticism (capital "G") and therefore nauseating and unacceptable to him.[41] Unethical bigotry, religious bigotry — take your pick. Neither should be acceptable as the premise for critiquing someone's work. It's sad that, as someone once said, progress in a field of study happens funeral by funeral.

I think it's kind to identify the scholars and publishers involved in the events documented here as "biased" and "unprofessional." Less kind, but more accurate, would be to say they are malicious, ignorant, contemptuous, and dishonest. It should be enough to point out, in a more clinical fashion, that they have failed to abide by the foundational principles of their professions. They have violated science.

My own place in this debacle is that of the outsider who made the breakthrough discoveries and sought a fair hearing among the appropriate scholars. There have been hearings, but they've been anything but fair. I have persistently and patiently asked the citizens of Maya Studies to abide by the principles of scholarship and science. One being that errors published in peer-reviewed journals and books must be acknowledged and corrected. In only one instance of the many examples presented here did a scholar freely acknowledge an error, or oversight, and that was Coggins' brief comment, that she had apparently overlooked how my work anticipated her own. The mere silent removal of offending statements, as in the

---

[41] Restall & Solari (2011:121) concur with Aveni on his Gnosticism construct, failing to recognize that his biased critique was unethically rooted in religious bigotry — violating one of the first principles of academic and scientific professionalism. My review of Restall & Solari's book is in Appendix 1; see link on p. 156.

Morrison example, is not the same as acknowledging an error, and Aveni's final comment was grudgingly mumbled after much persistence.

To summarize, the Aveni episode developed from my emails to Aveni in 2013 and 2014, pointing out to him some errors in his book and his other assessments of my work. After an initial exchange, he then ignored my several follow-up emails through October of 2014. Thus followed my contact with his publisher, the University Press of Colorado, requesting that they acknowledge several errors. I had carefully selected seven, for their unambiguous nature. In their process of assessment, my complaint was forwarded to Aveni and their "valued advisors." All of them — everyone involved — denied that the errors could reliably be determined to be errors (accept for the previously acknowledged error in his assessment of Grofe's work). So we couldn't even get to the stage of *correcting* the errors. I was told to go fix the problem myself, elsewhere. But then, months later, the revised and allegedly "corrected" eBook edition of Aveni's 2009 book was released. An opportunity had arisen to register at least a few corrections — perhaps just the simple test error I'd included in my complaint (the publication year of the McKenna brother's book). But that opportunity was not taken because, of course, *none* of my purported errors were agreed to actually be errors.

Well, the University Press of Colorado is a member of the AAUP (the Association of American University Presses). So, I next went to the committee within this agency, which is charged with overseeing and maintaining the academic standards of their member presses. I confirmed that their process is to independently and objectively assess any complaint of malfeasance against one of their members. Basic standards of peer-reviewed academic publishing must be upheld. After several months I was given a curt, 70-word reply *not* from the committee chairperson (Leila Salisbury) but from the AAUP central office, telling me that no disciplinary action would be taken. Like the Morrison example, they were refusing to acknowledge that any of the errors were actually errors.

Furthermore, they refused to admit that the University Press of Colorado had violated any policies or standards (such as error recognition and correction). No need to be specific, they were simply not going to take any "disciplinary" action. This indicates that the entire chain of culpability, from scholar to publisher to the supervising policy agency, is functionally broken. My subsequent emails and direct queries were evaded. However, the chess game was not over. I emailed Aveni directly and pointed out that he had stated the Izapa ballcourt alignment differently in his publication of 2000, compared to his book of 2009. There was a quite large 48° difference in the stated orientations. I asked him which one was correct. He said the earlier 2000 publication was correct. Thus, the 2009 statement in his book, as I had pointed out in my complaint, was *incorrect* — in his word, a "mistake." Here we have Aveni admitting, after the decision of the AAUP central office *to do nothing*, that he committed an error. But here we have to picture the devious criminal in a 1950s B-movie, gleefully admitting to a crime after an innocent verdict was given and court adjourned — *you can't get me now, as that would be double jeopardy.*

Morrison's errors and denigrating slanders, in his presentations and on the NASA website between 2009 and the end of 2012, were not fully corrected. After a series of emails, calls, baffles, and delays, three months passed and only the two statements on his blog were removed, without further comment or acknowledgment that they were wrong. Meanwhile, I emailed Morrison and he briefly responded. I then asked him to address the numerous and more egregious statements in his presentations, which are on Youtube as well as the websites for the scholarly venues that hosted him — one even being linked to his biography on Wikipedia (as an example of his anti-2012 work). More phone calls and emails, with no response. I had appealed to the NASA Office of Communications to get the first part accomplished, because there was a stated policy that NASA scientists must communicate honestly and promptly correct errors. That rule applied to Morrison's blog on the NASA website, but it also applies to anything Morrison said in his educational and public outreach efforts, acting in his capacity as a NASA scientist. Thus, the other examples I brought up to him also apply. He is currently in violation of the stated Communications Policy of his employer, NASA.

I might take it to a higher level and file my complaint with NASA's office of investigations. But we've seen what happened to climate scientist James Hansen when he did this — years of delays, grief, increased character assassinations, intimidation tactics and murder threats, court appearances, filings and counter-filings. Morrison, known to his colleagues as Dr Doom, represents this ugly, despicable, dishonest and corrupt side of NASA "science." His just desserts is that he has to look at himself in the mirror.

Ed Krupp's behavior is much like Morrison's, with whom he shared at least one 2012 conference and panel: A cowardly retreat when I asked him to acknowledge and address his mistakes — most of which pertained to my work. Specifically, he took a truncated quote from my 1998 book out-of-context to affect an appearance, to his audiences, that I believed 2012 meant doomsday to the Maya, and that I was mainly to blame for the 2012 mess. His characterizations of me and my work are so off the wall and totally false that they can only mean one thing — that he never read much of anything I ever wrote. The alternative possibility is that he was consciously malicious, and sought to deceive his audiences regarding the nature of my work.

John Hoopes, a clever wolf in sheep's clothing, was demonstrably exposed in my *Zeitschrift für Anomalistik* essay, but he refused to admit to even one of the easily demonstrated errors, including his misunderstanding of who originated and first used the "2012 phenomenon" phrase. The bad behavior and hack-job scholarship of Hoopes is so consistent and prevalent over the years that I can only provide a few highlights in my summary here. He crafted a category called "Mayanism," appropriating a term previously used in a proactive sense by other scholars (I exposed this in my 2009 book). He has given his students assignments to work on Wikipedia entries, which appears to be a form of "sock puppetry" — a forbidden practice in which you enlist others to hijack certain entries and skew the context and narrative toward your own biased position. He was, as he himself states, actively designing entries like "pseudoarchaeology", "Mayanism," and "the 2012 phenomenon". Curiously, nowhere is it stated in "the 2012 phenomenon" entry that Geoff Stray was the first to use the phrase, in 2002. Other errors abound, such as crediting Munro Edmonson with my Dark Rift identification of the galactic alignment, no doubt courtesy of Hoopes. This was the one important criterion that I've emphasized through the years which separates my work from what came before. I showed that the era-2012 alignment utilized concepts in the Maya Creation Mythology, and opened the door on legitimizing 2012 as a valid topic of rational investigation. The ballgame symbolism followed, and my work at Izapa.

We could thus, thanks to my work, explore evidence and data that could help us understand the 2012 astronomy and ideology — what the ancient Maya thought about 2012. But instead Hoopes worked hard to invent a narrative in which my work was just a seamless extension of an invented 2012 mythology, boosted by 1960s psychedelic culture, drug-taking hippies, and cultic Theosophists (in collusion with Nazis). Hoopes's construct works to the extent that it is totally superficial and treats my work with haughty disdain. He doesn't have to engage the originality, depth, insights, arguments, and evidence I brought to bear on my reconstruction work. He just has to say that the Nazi swastika symbolizes World Ages, that I was friends with McKenna or did an event where Argüelles also spoke, or that we all used the word "galactic" — never mind that McKenna and Argüelles weren't even concerned with reconstructing what the ancient Maya thought of 2012. They were coming at it from completely different perspectives, from that of a visionary model-maker decoding the I Ching (McKenna) or a game-playing myth-maker claiming to be the voice of a departed Maya king (Argüelles).

More flagrant violations of academic standards were committed by Hoopes, with the help of his publisher and friend John B. Carlson. Hoopes asserted that I once worked as an astrologer (false) and that the galactic alignment was basically astrology (false). Thus, a clever association with astrological "pseudoscience" could be asserted, based on these false and unsupported statements. He also said my work was based in the work of Dane Rudhyar (false, and Rudhyar never spoke of the galactic alignment), which was tantamount to an accusation of plagiarism since I represent my 2012 alignment reconstruction as being unprecedented and I don't cite Rudhyar for any of it. These egregious, unsupported, contemptuous slanders were published in a peer-reviewed journal — Carlson's *Archaeoastronomy Journal*. My attempts to speak of the matter with Hoopes, Carlson, and the publisher, the University of Texas Press, were met with ignorance, evasions, and denial. Carlson defended Hoopes's words, even though they were totally unsupported and negatively misrepresented a living author.

Hoopes's bad scholarship continued through several articles, in both peer-reviewed journals and the popular press (e.g., the *Fortean Times*). His collaboration with religions scholar Kevin Whitesides, published in a German journal in 2012, is where I had to draw the line. I succeeded in getting my corrective review of their article published in the same journal, and they were invited to respond. They wiggled around admitting to any of the factual errors, even though Whitesides had admitted to one in a previous email to me. Whitesides constructed a critique that my ideas proceeded from universal "archetypal" intuitions rather than evidence-based scholarly work that engaged with debate, defense, and discussion. Do I need to point out the absurdity of such a construct? My work is an interdisciplinary synthesis of evidence, which is clear in my 1998 book. I've been actively and persistently seeking discussion on these ideas for over two decades. And again, like

Aveni before him, Whitesides cited my 1998 book as evidence for his statements, where nothing of the sort can be found to support his assertions. I suspect Whitesides used this same critique of my work in his Masters dissertation,[42] awarded to him in 2012, and if he admitted to it being in error his degree might come under review. Well, science needs people who do science, not character assassins, so perhaps his degree should be revoked.

John B. Carlson is a cagey and uncommunicative guy. He exhibits a consistent pattern of evasion and then, late in the game, he echoes my idea of deity sacrifice and renewal in 2012. His statements in his 2010 Robbins Museum lecture about my knowledge base (or lack thereof) is demonstrably false. When I called him on this he retreated into isolation, speaking by proxy through his friends at the Robbins Museum. As the chief editor of a peer-reviewed journal he is also guilty of publishing and then defending Hoopes's unsupported assertions about my background and ideas. He is unable to engage in cordial discussion and is unwilling to admit when he has misjudged my work, and what my position is on Maya astronomy and 2012 — which inspired and anticipated the similar fact-based, reasonable ideas noted by other scholars (e.g., Michael Grofe's precessional work) that Carlson has published in his journal.

Susan Milbrath, the co-editor of an important anthology on Mesoamerican astronomy published in 2015, refused to address my query and instead obsessed on her misperception that I was accusing her authors of plagiarism — which I had explicitly stated, in my first email to her, that I wasn't assuming. Years earlier, I never had a full conversation with her about her inaccurate assumption regarding precision in the galactic alignment astronomy. Our earliest exchanges were suppose to be predicated on her familiarity with my work, after she received a free review copy of my *Maya Cosmogenesis 2012* book in 1997, but apparently she wasn't well apprised of its contents and appears to have simply disregarded it, even though many of her own findings about Maya astronomy followed similar lines of citation and argument (e.g., see Milbrath 1999 compared to Jenkins 1998). The recent exchange with Milbrath hit the wall with my attempt to point out the contradiction in how my work was echoed in two chapters of the anthology she edited, compared to my placement in Krupp's "End Times Follies" holding pen. (Krupp wrote the preface for her anthology.) If scholars can't look at the facts of a situation, even if they are unsettling and indict their own oversights, then, well, "Houston we have a problem."

I reflect back on my 30+ years of engagement with the Maya people, Maya culture past and present, the news media, and Maya Studies and see how many arcs and cycles in my life are now completing, full circle. I'd like to say I feel that, with the completion of this project, closure has finally been reached. But I'm beginning to feel that I'll never have closure with my 2012 work. As one can see by reading this book, all the examples discussed are left dangling and largely unresolved, because scholars and their publishers were evasive and dishonest. It's hard to say if they are

---

[42] He claims he didn't (email of February 12, 2016).

cowards, clueless, arrogant, or just plain careless. Probably all of these in varying degrees.

The purveyors of major scientific and academic breakthroughs rarely get to see their work accepted within their lifetimes. I will probably die before it gets a fair hearing and gets accurately portrayed and assessed. Most likely, the original work I did and the discoveries I first articulated will become subsumed into the consensus viewpoint in Maya Studies — credited to others while historians conveniently misremember who said what first. A younger generation of malicious debunkers following in the mould of a Hoopes or Krupp will just continue maintaining the false and denigrating narrative, crafted mainly by Hoopes, of my purported "associations", "inspirations" and "influences."

The events I document here are disappointing, and must be embarrassing to the scholars and institutions involved, but I document them in the hope that someone, somewhere, sometime, will see the injustice and corruption that has afflicted Maya Studies and has prevented an important idea from being assimilated (that of 2012 being an intentional forward calculation to a rare alignment within the precession of the equinoxes). I suppose the bad behavior follows the model enunciated by Thomas Kuhn (following Schopenhauer), the three-part process by which new discoveries in a field of study get integrated: 1) The breakthroughs presented by an outsider are ignored; 2) the new ideas and the messenger of them are violently attacked and criticized; 3) the breakthrough ideas are accepted by the status quo as if they had been known all along. There's no guarantee that the maligned innovator gets any satisfaction, recognition, or justice.

I can't do anything more on this, since I've encountered the stonewalling of agencies and publishers that have no conscience, the evasions of dishonest scholars, and the robotic, oddly hollow, decisions of people who run those agencies and publishing ventures whose hands (and minds) are apparently tied — tied to serving corporate and legalistic dictates, rather than the principles of good science. The Ivory Tower is thus revealed as nothing but a house of cards, poisoned by its own internal malfeasance, ego politics, irrational evasions, and dishonesty. I wish I could end on a more positive note. These are dark days for those who still believe in honesty, open communication, conscience, and integrity. I asked scholars and scientists, their oversight agencies, policy committees, and academic publishers, to behave according to the basic policies and principles of their profession. About 95% of what transpired reveals an abject inability or unwillingness of them to do so. What would the report card show? Well, let's just say there are no shiny gold stars or smiley faces.

October 30, 2015, closure achieved in communication with
the AAUP's Executive Director, Peter Berkery.
Conclusion: **Checkmate.**

◇ ◇ ◇ ◇ ◇

And this was the conclusion of my study, completed in November of 2015. In 2016 the academic obfuscation machine launched another grenade into the understanding of 2012 and Maya precessional knowledge. Again, it was courtesy of Anthony Aveni and the University Press of Colorado. This time around I was able to use my hard-won checkmate achievement, a continued adherence to facts and evidence and rational argument, as well as a communication that I'd be willing to pursue the unresolved situation in the courts. In any case, half way through my renewed effort of 2016 the tide suddenly shifted and all the errors I pointed out were duly acknowledged. This long overdue and unprecedented breakthrough is described in the following addendum.

By way of introducing the Addendum, here is an informed reader's comment on my review of Aveni's 2016 book:

"Your review of Aveni's 2016 book points to a general lack on the part of Aveni to pay much attention to data and evidence put forth by known Mayan scholars. Instead it appears he has taken out of context, or in most cases falsely cited, you and others (including himself!) with what he seems to think is evidence to support his claims. In science that is a difficult data set to overcome, as you well know. Authors like Aveni who manipulate data to make one think those pieces of data can then be explained as evidence to back a particular stated view of a topic are the scourge of real science. Science by its very nature must look at as much data as possible and then draw reasonable assertions without prejudice. Selecting only those bits and pieces that are really misrepresentations of other scientists' intentions is clearly a violation of the scientific/scholarly endeavor, which scientists are necessarily held to.

In the case of Aveni it appears he has violated most of the known rules of both citing other authors in the field, and attacks those with whom he disagrees without having any substantiated basis for his arguments. Pointing out that Aveni didn't have the ballcourt orientation correct, and was off by 48 degrees, shows he either didn't check his own data or used the data of others without doing the observations and measurements himself. This would lead one to be skeptical of other purported facts set forth by Aveni. What I read in your critique is an author (Aveni) who has constructed a work(s) laden with factual and purposeful misrepresentations of other scholarly works. It is obviously a way to sell books and create a myth that he (Aveni) is the singular expert in the field of Mayan Studies."

—William Crandall, science educator

Addendum 2016:
Correcting Aveni's 2016 Book *Apocalyptic Anxiety*

In June of 2016 I was astonished to discover that Aveni had published another book related to 2012, again with the University Press of Colorado. It, again, contained many errors (some of the same ones from his earlier book on 2012). I went through the same process of requesting the errors be corrected. It looked like the same old tricks were going to be employed, so I called Pratt and said that these continuing problems, and the way they were being handled, were unacceptable and I was prepared to file a cease-and-desist order. This seemed to change the game, and Peter Berkery (who was cc'd on my new effort) may have had some words with Pratt. The result was that Aveni agreed to all the needed corrections. By October the remains of the first printing were pulped while the second printing and the eBook were revised and released. It was another grueling effort, one that I couldn't believe was necessary. I produced a lengthy analysis and review of Aveni's book (see www.thecenterfor2012studies.com/Review-of-Aveni6-2016.pdf), but eventually I boiled it down to this review-essay, which has an update at the end.

Hand-wringing in Maya Studies:
Approved Corrections to the Second Printing of
Anthony Aveni's *Apocalyptic Anxiety*

John Major Jenkins. © October 5, 2016.

This new book by Anthony Aveni (*Apocalyptic Anxiety*, May, 2016, University Press of Colorado) demonstrates that the topic of "2012" is still relevant and subject to treatment by an academic scholar in Maya Studies. Aveni, primarily known as a pioneer of Native American archaeoastronomy, explicitly uses the 2012 episode in our recent history as the closing bookend of his treatment, which he compares to the Millerite hysteria of the 1840s. Aveni locks these two episodes together in a 168-year-long tale of America's obsession with apocalypse. Like many other academic books on 2012,[1] Aveni doesn't recognize the efforts of researchers who have worked to reconstruct what the ancient Maya thought about 2012. Instead, the entire topic is framed as millennial hysteria, anti-modernism, and New Age fantasy.

Focusing on the underinformed mass-appeal response to this misunderstood event or topic is certainly one way to look at 2012, but it ignores what should be of more interest, and relevance, to an archaeoastronomer and Maya Studies scholar like Anthony Aveni. Which is: What did the Maya think about it? Why does the most widely accepted correlation place the 2012 period-ending date on an

---

[1] This includes virtually all of the "academic" books on 2012, which in varying degrees (usually almost *completely*) disregarded research and findings that treated 2012 as a valid artifact of ancient Maya thought. These incomplete books include Stuart (2011), Van Stone (2010), Restall & Solari (2011), Aveni (2009) and the 9th edition of Michael Coe's *The Maya* (1966), co-revised with Stephen Houston (2015).

astronomical solstice? Based on his previous 2012 book (*2012: The End of Time*, 2009) and his other statements in presentations at academic venues, for him it is very unlikely that the ancient Maya thought anything much at all about 2012. In this position he is allied with colleagues David Stuart, Stephen Houston, and now Michael Coe.[2] Never mind the growing body of research done by myself and five or six other Maya scholars, and never mind the direct communications I've had with Aveni about this work. That has no place in his book.

And yet there I am, referenced and treated in the final part of his book and a few other places throughout. Recognizing this contradiction (what my work is actually about versus how Aveni misrepresents it) will help us understand a major flaw in Aveni's book, which signifies a widespread general flaw regarding how many Maya scholars disdainfully mishandled 2012. Part IV includes chapters on "Galactic Wisdom" and the "Perennial Philosophy" that address important aspects of my work. I am included but am introduced as following "in Argüelles's footsteps"[3] (202) and as a 2012 "prophet" who makes "prognostications," not as someone who has proposed and argued for an unprecedented reconstruction of ancient Maya precessional cosmology related to 2012, in articles, books, and presentations given at popular as well as academic venues for over twenty years. I am force-fit into a narrative in which 2012 is/was a modern "invented mythology,"[4] and Aveni is aided in this effort by anti-2012 critic John Hoopes. We'll see how all this is cleverly crafted in Aveni's book, and is effectuated by committing a dozen factual errors of various sorts.

A longer review-critique of Aveni's book is necessary because I have already successfully facilitated a dozen corrections in Aveni's book, after communicating with Aveni and his publisher, the University Press of Colorado. I assumed and followed the standard protocols offered by every reputable academic publisher, regarding errata, and Aveni has now, as of mid-August 2016, acknowledged and corrected all the errors.[5] However, this unprecedented victory came after many

---

[2] After adamantly maintaining his original doomsday interpretation for some 49 years (compare Coe 1966 and Coe 2011; see also Jenkins 2015), Coe deleted his "Armageddon" interpretation of 2012 in the 9th edition of his book *The Maya* (1966), which was co-revised with Stephen Houston and released in June 2015. Addressing the evidence from two 2012 inscriptions, they incredibly claim they are unremarkable: they "tend, if anything, to be rather dull." (Coe & Houston 2015:250).

[3] With his "Dreamspell" in the early 1990s, Argüelles inspired about the closest thing to a cult that the 2012 Phenomenon produced, declaring himself the voice of a discarnate Maya king. I called out these shenanigans myself, critiquing his Dreamspell calendar system (*Tzolkin* 1992/1994), and received a lot of flack for doing so. For Aveni to imagine that I was following in Argüelles's footsteps demonstrates a lack of knowledge of actual events, unsupported by the published facts.

[4] Aveni adopts this concept from Hoopes (2011) and, through Hoopes, from Hammer (2001) and Hammer & Lewis (2008). In regard to the Perennial Philosophy, the "invented mythology" critique is flawed due to a fundamental misunderstanding about what the Perennial Philosophy is; see discussion below.

[5] The circumstances of how this came to pass is a long story, and relates to my earlier request that many factual errors in Aveni's 2009 book be corrected. After ten months of effort, Aveni, his publisher (the University Press of Colorado), and the AAUP denied the

emails, a phone call, and two months of effort, some three months after the book's release, by which time the first printing was sold out. Consequently, according to the World Catalog online some 135 libraries around the world (mostly university, college, museum, and technical school libraries) are holders of the error-containing first printing. A useful primary purpose of this review is to report the corrections and to alert holders of the first printing to this misleading situation, as a recall of the first printing was not offered by Aveni's publisher. I will also briefly discuss the implications of these errors for Aveni's overall interpretations in his book, which short of a recall and complete rewrite could not be addressed in line-item editing. Although the second printing and the eBook will record the basic factual corrections, there will be logical contradictions in Aveni's text because some of his interpretations and reasoning cannot be maintained in light of the errors that have been acknowledged and corrected.

Much of the first two-thirds of Aveni's book runs through previous episodes of end-of-the-world hysteria in American history. It's a standard litany that draws the typical narrative of dubious fringe-cults from books like Horowitz's *Occult America*. All of it has the feel of setting up Part IV, where Aveni deals with what he sees as the most recent hysterical apocalypse episode, which is "2012." Given the space limitations in my review, as well as my own intimate knowledge of and involvement with this particular episode, stretching over some thirty years, I will focus on this material. This is, after all, where Aveni brings together his arguments and offers his conclusions, based on the evidence and citations he marshals. It is also, curiously, where all of the errors are concentrated and thus where a corrective review-critique should focus.

As mentioned, marketplace and media hysteria was an unfortunate and large component of the "2012 phenomenon."[6] Before and during these distractions, I proposed and maintained a serious rational treatment of what the ancient Maya likely thought about 2012. Most scholars, however, chose to merely engage in a debunking of the media mess, rather than pursue their own 2012 research into Maya calendrics, astronomy, 2012 inscriptions, the Creation Mythology and other Maya traditions. This style of 2012 critique centrally involves Aveni, who was the only Maya scholar who published a book-length treatment of the topic with a peer-review university press (Aveni 2009). His 2009 book was also the *earliest* full-

---

errors and/or refused to take action. However, my subsequent discussion with Peter Berkery, the Executive Director of the AAUP, allowed me to share Aveni's admission to one of the errors, exposing a flawed assessment process, and a checkmate situation for Berkery. I believe that this precedent has, this time around, forced scholar, publisher, and supervisory agency to take seriously their academic obligations and avoid legal action or professional embarrassment.

[6] Contrary to statements made by Robert Sitler (2010), Whitesides & Hoopes (2012), Whitesides (2015), and other scholars, and despite my explicit published corrections (e.g., Jenkins 2014a), Sitler did not coin and first use this phrase in his 2006 *Nova Religio* essay. Rather, it was demonstrably used as early as 2002 by Geoff Stray, on his extensive Diagnosis2012 website, as well as in his 2005 book *Beyond 2012*. It was frequently used by Stray, myself, and author Jonathan Zap for years prior to 2006.

scale book treating the topic, and subsequent scholars often have cited his book as a viable assessment of 2012 authors and ideas.[7]

A major problem with Aveni's overall approach, in both of his books, is the lack of distinction between evidence-based reconstruction efforts, such as my own, and the marketplace mess that usually had little to do with an accurate presentation of Maya traditions. The paradox which underscored Aveni's misunderstanding of the nature of my work is that he falsely conflated me with the marketplace mess, used his perception of my religion against me, and his loose and unqualified associative statements sometimes insinuated that I was a "Y12er," one of those doomsday prophets. Aveni's misinformed convictions are largely maintained in his recent book. I pointed out the factual errors to him and his university press publisher in June of 2016. The process of error assessment and correction I initiated was identical to what I attempted in early 2015, regarding his 2009 book, which also contains many factual errors. That ten-month-long effort resulted in a denial of the demonstrable errors by Aveni, his publisher, and the AAUP (the Association of American University Presses, which ratifies and supervises their member presses). All involved are bound by academic and scientific policies of error acknowledgment and correction, but all involved did not abide by those policies.

This time around, to his credit, Aveni has acknowledged and corrected all of the errors I indicated in his recent book, including a few that were also present in his previous book. The reason why the process was smoother this time around is probably rooted in two things: an email exchange I had late last year with Peter Berkery, the Executive Director of the AAUP, and a phone conversation I had this year with Darrin Pratt, the Director of Aveni's university press publisher, when I learned that they would simply rely on Aveni's own assessment of the "alleged" errors, rather than initiate their own objective assessment. It was surprising to learn what it takes to get these professional academic establishments to effectively perform, and follow through with, what is after all a central mandate of scholarly writing and academic publishing.

With the errors confirmed, Aveni's publisher promised that the corrections would be added to the second printing and the eBook (see my Addendum after the bibliography for an update). Nevertheless, as mentioned the first printing was already sold out by the time the errors were finally honestly dealt with (in mid-August), with 135 copies already on the shelves of libraries around the world. So, the damage had already been done, as the saying goes. An offer to add an errata sheet to the remaining stock of the first printing was no longer on the table, because the first-printing stock was almost sold out. For anyone, or for any scientific institution or publishing venture that is concerned with the published record being accurate, it will therefore be valuable to document the errors in this review. It is a breakdown of academic publishing when such systemic errors pass fact-checking scrutiny at a very high level of scholarship and academic publishing, with the result

---

[7] E.g., Krupp (2015), Restall & Solari (2010), Van Stone (2010), Whiteside & Hoopes (2012, 2014).

being a very misleading and factually flawed book going onto the library shelves, unimpeded and without being retracted.

Many of the twelve demonstrable errors in Aveni's new book have severe repercussions for accurately understanding the ideas found in the "2012 episode" and my role in reconstructing ancient Maya traditions that relate to 2012 — particularly, Maya astronomy and Creation Myth beliefs. To be concise up front, I provide below my brief synopsis of the errors, which I sent to the author and his publisher in early July, 2016:

A Quick Summary of Selected Errors in Aveni's Book

Error 1. The assertion and mitigating construct that my perennial philosophy ideas drew heavily from the work of Mircea Eliade is false, is not evident in my work, and is not supported by the source Aveni cites for support (Hoopes 2011).

Errors 2 and 3. An explicit quoted phrase and paraphrased material from my 2009 book (*The 2012 Story*, Tarcher/Penguin Books) are incorrectly credited to Olav Hammer.   Hammer concurs on my observations and corrections. Also: Aveni cites to a non-existing title that Hammer never wrote.

Error 4. (Actually, five related errors are packaged together here.) The Argüelles citation errors (a total of three), the un-indicated missing phrase from one of the quotes, and Aveni's demonstrably incorrect unqualified assertion, in square brackets, regarding Argüelles's intended meaning. The effect of these various errors is the imputation that Argüelles was aware, in 1975, of a central feature in my unprecedented reconstruction work on Maya astronomy, which I first published in 1994.

Error 5. The mis-portrayal of my work as "following" in "Argüelles's footsteps," in contradiction to the well-known published facts of my disagreements with, exposés of, and critiques of Argüelles's ideas, beginning early in my writing career (Jenkins 1992/1994).

Errors 6 and 7. My reconstruction work at Izapa does not offer "prognostications," as Aveni states. It is an interdisciplinary reconstruction based on evidence at the site from the fields of archaeoastronomy, calendrics, environmental determinants, anthropology, and iconography. Finally, the Izapa ballcourt alignment needs to be correctly stated, for Aveni states it, as he did in his 2009 book, 48° in error.

Error(s) 8. A related set of errors in the Index confuses me with the Rapture/Left Behind author of Christian doomsday fiction, Jerry B. Jenkins.

I also sent a more detailed treatment of these errors, with pointers to the supporting evidence. Some of these details are worth sharing, in order to show the unambiguous and non-negotiable nature of the errors. A major error involves

Aveni's direct critique of my work, which he bases on his incorrect assertion that I "drew heavily on the work" of Mircea Eliade (202). As support for his assertion, Aveni cites an article by John Hoopes,[8] where no statement of the kind can be found. (Aveni thanks Hoopes in his Acknowledgements, whose critiques of 2012 he states "directly influenced" his book.) In addition, a survey of my four primary books between 1998 and 2009 shows that the presence of Eliade is practically zero. In fact, there are no discussions, citations, or quotations from Eliade in any of them. My World Tree (*axis mundi*) knowledge began with readings in Hinduism, Carl Jung, world religions, and Finnish mythology in the early 1980s. In my book *Galactic Alignment* (2002), which deals most extensively with the Perennial Philosophy, I list a dozen Perennial Philosophers who have been important to me and Eliade does not appear, anywhere. Asserting (falsely) that my ideas drew heavily from Eliade was clearly an important critical point for Aveni to establish, because he had already criticized Eliade as a nostalgic anti-modernist "perennialist" who invented phrases like *axis mundi*, who thereby fed a "fantasy-loving, gullible, popular culture" (Aveni 2016:185), and whose work inspired a post-modernist trend toward anti-intellectualism (Aveni 2016:184).

These mistakes are breaches of sound scholarship, and they are just the tip of the iceberg. Another error was committed by Aveni's publisher, which reflected Aveni's negative characterizations of me. In the Index to Aveni's book my name is not found, but the pages in the book where I am mentioned and my work is discussed are listed under Jerry Jenkins, who is the Rapture/Apocalypse "Left Behind" Christian fiction author. Under his name, the pages that refer *to my work* (pp. 202-203) are delineated with the sub-heading "Maya end of world." So, the incorrect identification of my work as "Maya end of world" information that rubs shoulders with fictional Rapture-awaiting Satan smashers, is accomplished.

Aveni introduces me as "following the cosmic road in Argüelles's footsteps" (202). On the contrary, my critiques in the early 1990s of José Argüelles's ideas about the Maya calendar exposed how they do not accurately reflect Maya concepts and the traditional placement of the 260-day calendar. Based on my concern for accurately portraying the Maya calendar tradition, and trying to educate those who didn't share this value, I had an oppositional stance to Argüelles's attempts to craft a new dispensation that had cultic overtones, a new mystical calendar only loosely based on the authentic Maya calendar.[9] As such, my work pioneered the critique of the 2012 Phenomenon in the early 1990s, long before scholars such as Robert Sitler and John Hoopes entered the picture.[10] The wide gulf of difference between

---

[8] "Mayanism Comes of (New) Age" in Gelfer (2011). See http://update2012.com/Gelferanthology.pdf.

[9] See, e.g., Jenkins (1992/1994) and "Following Dreamspell": http://alignment2012.com/following.html.

[10] In his practice of selective academic ignorance, Hoopes has diligently avoided acknowledging my early efforts in the critique of the 2012 Phenomenon. In his many narratives about the "2012 Phenomenon" he also refuses to acknowledge that it was a phrase used by myself and Geoff Stray years prior to Sitler's *Nova Religio* essay in 2006, who he and Kevin Whitesides (2012) credited with coining the phrase. See my review-essay here: http://update2012.com/Jenkins-Zeitschrift-fur-Anomalistik-1-2014.pdf.

Argüelles's "galactic synchronization" concept and the astronomical facts of the galactic alignment is clearly explained in my work,[11] which Aveni ignores, instead suggesting that I have merely carried on Argüelles' work — a flawed and baseless position that Aveni has repeatedly asserted for years,[12] despite my corrections. Although he has now acknowledged and corrected these problems, Aveni's previous critiques show an undiscerning blending of terms and concepts, with little effort to understand his subject or to accurately cite my work on the matter.

As mentioned, Aveni writes in his Acknowledgements (xv), that he was "directly influenced" by the work of John Hoopes, who has constructed an anti-2012 critique of "eclectic and non-codified" New Age ideas that revolve around 2012. He calls it "Mayanism."[13] Curiously, an early working sub-title for Aveni's book appears to have been "From Millerism to Mayanism." Hoopes's Mayanism has been repeatedly debunked as a flawed construct,[14] on grounds of semantics and *inverting the meaning of the term* as it was used by anthropologists in the 1990s. The point of Hoopes's efforts was to frame "2012" as an invented mythology, an *invented sacred tradition*. In doing so he indicts my work because I have identified, within Maya tradition, a World Age doctrine of period-ending renewal that points to 2012. You see, "renewal" means a New Era, a New Sun, a New Age, that was expected by the Maya in 2012. Since Hoopes's Mayanism relies on the "New Age" concept as a sure hallmark of the Mayanism heresy, then my work must be part of Mayanism. See how that works?[15] This rationale is malicious and fallacious. Hoopes indulges in this hostile fallacy and is unable to acknowledge my fact-based findings, which I addressed in my peer-reviewed essay of 2014:

> My usage [of the New Era *renewal* concept] was not derived from McKenna, Argüelles, Blavatsky, or the New Age movement, as some critics assert (e.g., Whitesides & Hoopes, 2012; Hoopes, 2011: 54). It arose from my investigation of the evidence at the site of Izapa. That the ancient Izapans and Maya (and other cultures) had a World Age doctrine in which world renewal occurs at specific intervals should not be obviated by the fact that such ideas are superficially echoed in the modern New Age marketplace. Critics need to apply discernment to recognize the distinction. (Jenkins 2014a: 56)[16]

---

[11] For example, Jenkins (2009:101-102) and www.alignment2012.com/5misconceptions.html

[12] His Colgate presentation of early 2012 and his Penn Museum talk of Dec. 2012, both on Youtube.

[13] See Wikipedia entry, especially the Talk pages, where the validity of the entry is challenged without any effective rebuttal. The page was called to be deleted. See also my critiques in Jenkins 2009 and 2014.

[14] See, e.g., http://www.alignment2012.com/Mayanism-John-Hoopes.pdf.

[15] And yet many other scholars have expressed this same basic notion about 2012, using terms like "renewal", "new cycle," "era transition." For example, in February 2005 Barbara Tedlock stated: "2012 is an important date in the Maya long count. ... It's just the end of one era; the beginning of another. It is not the end of the world" (B. Tedlock 2005:42).

[16] "The Coining of the Realm (of the 2012 Phenomenon)" in *Zeitschrift für Anomalistik* Band 14 (2014), No. 1: www.update2012.com/Jenkins-Zeitschrift-fur-Anomalistik-1-2014.pdf.

Great cognitive dissonance is Hoopes's lot because my interpretation that "worldrenewal requires deity sacrifice in 2012" is completely reasonable, is in accord with known concepts of Maya period-ending ceremonies, and was echoed (albeit late in the game) by his friend John B. Carlson.[17]

During our email exchanges Aveni received the essays I sent him or summarized for him in mid-2014, including my exposé of Hoopes's Mayanism and my peer-reviewed *Zeitschrift für Anomalistik* piece. Nevertheless, he must have ignored my peer-reviewed scholarship and pointed comments. Consequently, the important distinction between reconstructing a Maya concept of "New Era" renewal at a big calendrical period-ending and various free-form "New Age" expressions in the marketplace was lost on Aveni. For some reason he chose to ignore my own statements and instead employed Hoopes's flawed and hostile constructs, possibly because they would bolster his own dismissive attitude toward my work.

Hoopes also guided Aveni to the book called *Claiming Knowledge* (2001) by Olav Hammer, which critiqued the Perennial Philosophy as part of "constructing a tradition" (Hammer 2001:155, 170-176). This interpretation assumes the Perennial Philosophy to be a man-made system, which is counter to what Perennial Philosophers themselves describe: "The Primordial Tradition or *sophia perennis* is of supra-human origin and is in no sense a product or evolute of human thought."[18] But Hammer's inversion (which is equivalent — not *in the meaning* but *by analogy* — to asserting that "atheists believe in God") was useful for Aveni because Aveni's Chapter 11 is titled "2012 and the Perennial Philosophy" and, there, he sought to critique the topic. This chapter would thus supposedly explore my proposal and long-argued position that the Maya "ideology" (or Creation Myth "teaching") that is associated with period endings like 2012 is also found in the Perennial Philosophy. Oddly, however, my name and my work are not found or cited anywhere in Aveni's Chapter 11. In addition, my central reason for making the connection between 2012 and the Perennial Philosophy — that **deity sacrifice is necessary for worldrenewal in 2012** — was never conveyed in Aveni's book.

Rather, it is in Chapter 12 (p. 202) that Aveni mentions my 2012/Perennial Philosophy proposal. Apparently trying to represent my thoughts on the matter, he proceeds to put together a series of truncated cherry-picked quotations from my 1998 and 2009 books, separated by ellipses, to create some disjointed paraphrases. It's a blatant and disconcerting display of academic fiddling with source material.

---

[17] Carlson 2011. My work is based on my reading of the Maya Creation Mythology on the monuments of the Izapan ballcourt. However much my interdisciplinary methodology may be criticized, Hoopes has never explained how my 2012 ideas were echoed much later by Carlson and other scholars, who came late to the rational treatment of 2012 as a valid artifact of ancient Maya thought. For example, see essays by Carlson and Callaway in the Oxford Archaeoastronomy IX papers. See http://thecenterfor2012studies.com/2012center-note10.pdf and http://www.cambridge.org/us/academic/subjects/astronomy/astronomy-general/archaeoastronomy-and-ethnoastronomy-iau-s278-building-bridges-between-cultures?format=HB. Not to mention the serious treatment of the galactic alignment by Grofe (2011a, 2011b) and MacLeod & Van Stone (2012).

[18] *The Betrayal of Tradition*, p. xii (2005, ed. H. Oldmeadow, World Wisdom Books).

And, in any case, the material he selected *does not* explain why I have proposed a connection between Maya concepts of 2012 and the Perennial Philosophy. That is found in my 2002 book, in numerous places in my 2009 book (e.g., Chapters 8 and 9; pp. 75, 228), and I summarized it *in eight words*, bolded in the previous paragraph above.

Aveni's use of Hammer's book *Claiming Knowledge* is problematic, for three reasons:

- Hammer inverts a basic premise of the Perennial Philosophy and frames it as a constructed tradition (the topic appears under the section "Constructing a Tradition" in his 2001 book).

- Hammer projects his own dualistic bias onto the non-duality of the Perennial Philosophy in order to explain how the *many* exoteric expressions are reconciled with an underlying *unity*.

- Hammer ignores the primary voices of the Perennial Philosophy, instead allowing it to be represented by the various cult figures within the Theosophical Movement and many other distorted derivations of Vedanta / Perennial Philosophy.

Hammer's later books and anthologies drop the Perennial Philosophy emphasis and focus on the secondary historical distortions propagated through the Theosophical Movement. This suggests that he began to understand that it was inappropriate to project the secondary distortions back onto the original inspiration. For example, in *The Invention of Sacred Tradition* (a title Aveni cites in a grand conclusion to his Chapter 11 critique of the Perennial Philosophy), Hammer & Lewis wrote: "Theosophy, a religious current with roots in the nineteenth century, claims to be an expression of perennial wisdom" 124-125). There's only one other reference to the "philosophia perennis" in that entire book, in an article on Sufism.

Hammer is the author who Aveni mistakenly credits (178, 233) with a specific quotation from my work, as well as with a paraphrase of my four-point summary of the Perennial Philosophy. In yet another incorrect citation performed by Aveni, he cites these to a book by Hammer titled *Philosophia Perennis*. But Hammer tells me he never produced a book or article with that title (p.c. June 2016). My literary forensics on Aveni's citation mess shows that he was intending to cite pages 172-173 and 175 from Hammer's *Claiming Knowledge* book, in the sub-section called "The Perennial Philosophy." Here, Aveni (178) summarizes Hammer's notion that Perennial Philosophers must believe that practitioners of the *exoteric* rites of a religion are self-deluded, because it is the inner *esoteric* symbolism that has the deeper, unified meaning. This is a ridiculous and unwarranted assessment. It bespeaks the Cartesian either-or dualism that many scholars are stuck in, when in fact Perennial Philosophers employ a non-dual understanding of the relationship between object and subject, exoteric and esoteric.[19] There is no *inherent conflict* between the Relative and the Absolute that requires a deceptive rationalization.

Nevertheless, Aveni uses Hammer's assessment, probably because he too is intellectually challenged by the concept of non-duality. Aveni asserts in his preface that "the two basic ways of knowing — reason versus revelation — are irreconcilable" (xiv). Take note that to *intellectually understand* the concept of non-duality does not require that one has had a revelation of God, or an initiation into Secret Holy Mysteries. Clearly, Aveni is hostile to the concept and employs Hammer's misleading "more radical" (Hammer, 173) notion that attempts to explain how Perennial Philosophers reconcile the many various exoteric religious expressions with an underlying unified source. But, as mentioned, Hammer's interpretation is fundamentally flawed as he projects his own either-or dualism into a non-dual worldview that does not fall prey to such a limited cognitive framework.

So, Aveni cites Hammer's pages 175 and **319** for an explicit quote and for a four-point summary of the Perennial Philosophy. These are not found anywhere in Hammer's book, and instead can be demonstrably traced to pages 292 and **319** in my book *The 2012 Story*. Aveni cites my book elsewhere in his book, and another paraphrase crafted by Aveni (178) closely reflects my words on page 290 of my book, under the heading "What is the Perennial Philosophy?" This occurs just before Aveni launches into "his" four-point paraphrase (178). Paraphrasing specific sentences from another author's work is Aveni's style, as we can see in how he paraphrases Hammer in two examples (pp. 177-78). (I will here remind the reader that Aveni did, ultimately, acknowledge and agree with my decipherment and correction of these citation errors.)

There are other fundamental problems with Aveni's assumptions. He confuses perfection and wholeness (10-11, 185). This is relevant to "wholeness" being an attainable goal of spiritual awakening whereas "perfection" is a Christian guilt-trip mandate that is basically impossible to achieve. Aveni adamantly holds to an anti-World Age bias, evident in his discussion of *Hamlet's Mill* (Chapter 9), and so he doesn't recognize recent Maya Studies scholarship that shows evidence for the ancient Maya being aware of the precession of the equinoxes (the two concepts go together in Maya thought). In fact, he doesn't even mention any of these new breakthroughs in his book.[20]

Next (pp. 204-205) we have a paragraph that is densely populated with errors and misleading assertions, all of which have the effect of distorting and misrepresenting my work at Izapa. Given my previous direct communications with Aveni and his academic publisher, in which I corrected several of these same errors in his 2009 book, it is difficult to avoid the impression that Aveni is, here, just repeating known errors. In any case, his repeated errors violate the principles of responsible science and scholarship. Let's take a look, for this provides another iconic example of what

---

[19] See, e.g., Nasr's *Knowledge and the Sacred* and Coomaraswamy's collected essays (Princeton, 1977).

[20] MacLeod (2008, 2012); Callaway (2011); Grofe (2011a, 2011b, 2012-2013); Jenkins (2010; 2011a; 2011b).

is so factually misleading about Aveni's book. Please note that it has nothing to do with his inability to understand "spirituality" or the Perennial Philosophy; it has to do with his repeated assertions of factual errors and maintaining a skewed, inaccurate, and incomplete portrayal of my work. A very distorted picture is presented regarding what my work at Izapa is about, which I have clearly presented in all three of my primary books (1998, 2002, 2009), presentations at the *Society for American Archaeology* (2010) and the *Institute of Maya Studies* (1997, 2011), as well as in various essays and presentations that I freely shared online.[21]

As part of what he inaccurately calls my "prognostications" (204)[22] he introduces Izapa as "early classic ruins" (no, its heyday was in the pre-Classic, before 100 AD). He states Izapa was peripheral (no, it was the most prominent central site of the Izapa-Soconusco Isthmian civilization). He states Izapa was "non-Maya." Although semantically true, Izapa contains some of the earliest depictions of the Maya Creation Myth (the Hero Twin story). A continuity into the Guatemala Highlands and with the iconography of the Classic Period site of Copan in Honduras (on the same important latitude as Izapa) is well-documented, which Aveni doesn't consider. As such, Izapa could be said to have pioneered central ideological traditions of the Classic Maya. This is to say nothing of the probable origin of the Maya calendars within the Izapan culture, which Aveni chides when I talk about it,[23] conveniently ignoring the fact that his colleague Prudence Rice came to concur with this same position (she wrote the Intro to his 2009 book). Perhaps the Izapans *became* the Maya, and thus were "early Maya"; certainly their ideas and traditions were adopted into Classic Maya civilization.

Aveni repeats his mistaken reading of the Izapan ballcourt alignment, which I was the first to publish in my 1996 *Izapa Cosmos* monograph and in my 1998 book *Maya Cosmogenesis 2012*. Aveni & Hartung published the Izapa ballcourt alignment, correctly, two years later, in 2000, as part of a general survey of Pacific Coast sites.[24] My priority on the publication of this information has clearly become a sore point for Aveni, and it is a central piece of evidence in my reconstruction of the cosmological interests of the Izapan skywatchers.[25] Fact: The Izapan ballcourt is aligned to the December solstice sunrise horizon. The direction of viewing, towards the sunrise, is confirmed by several factors, all of which Aveni neglects to acknowledge:

---

[21] Such as: http://alignment2012.com/monuments-Izapan-ballcourt.pdf.

[22] Aveni's use of loose and loaded lingo seems to be his specialty. There are many examples in his 2009 book that border on slander. One qualifies as bigotry, because he identified me as belonging to *the religion of Gnosticism* and then used that (I am a "Gnostic" New Ager, etc) when critiquing my scholarship.

[23] See his Penn Museum presentation of Dec. 2012: www.youtube.com/watch?v=4roz-DGShmc.

[24] Aveni & Hartung, 2000. "Water, Mountain, and Sky: The Evolution of Site Orientation in Southeastern Mesoamerica." In *Precious Greenstone, Precious Quetzal Feather*, ed. E. Q. Keber. Labyrinthos.

[25] Curiously, although at least a dozens scholars have published critiques of my 2012, none of them ever mention my Izapa ballcourt alignment discovery, which is an important centerpiece of my evidence.

1. The throne on the west end of the ballcourt has a head on its front face, facing the sunrise direction, and a person sitting on the throne would of necessity face the eastward sunrise.

2. Behind and on a rise above the throne, one finds six flat "seating stones," backed up against a wall of the temple mound to the west. The only direction of viewing for those who sat or stood on these flat stones is toward the east.

3. The westward direction is blocked by Mound 125a, which probably also had a wooden structure on the top, further blocking any unimpeded view of the westward sunset horizon from the ballcourt. This is unlike the *clear view* of the eastward sunrise horizon.

Despite all this, which is presented in *Maya Cosmogenesis 2012* (1998), a chapter in my *Galactic Alignment* book (2002), and is summarized in *The 2012 Story* (2009), Aveni states that I "discovered building alignments [at Izapa] with the **winter solstice sunset** position" (204, emphasis added). This incorrect statement repeats one of Aveni's mistakes in his 2009 book, where he stated that the Izapa ballcourt is aligned to the "**December solstice sunset**/June solstice sunrise direction" (Aveni 2009:54, emphasis added). These statements dislocate the factual orientation of the ballcourt, and so provide an orientation that is *48 degrees in error*. I had informed Aveni of this mistake in an email I sent him in mid-2014, also sending a mini-essay about it and other errors that I had just posted on my website.[26] My cover letter to him was dated 6/27/2014 and had the subject line "My review of your comments on the Izapan ballcourt alignment." The cover letter reads, in part: "I don't know if anyone ever pointed out to you several errors in your comments on the Izapan ballcourt, in your 2009 book. Notably, you wrote that the ballcourt alignment is to the December solstice sunSET and June solstice sunRISE. This is no doubt just a guffaw, but I think it would be important to correct."

To this, Aveni had no response (despite an ongoing email exchange). Some six months later, in early January of 2015, I filed a formal complaint with the University Press of Colorado, enumerating a half dozen or so factual errors in Aveni's 2009 book. His mistaken reporting of the ballcourt alignment was one of them. Aveni was sent my list of errors by Darrin Pratt, the press Director, whereupon he denied this and the other errors pertaining to my work. Later, in October I asked him to tell me which of his ballcourt orientations was correct, and which one was in error. As mentioned, he had published the correct orientation in his article with Hartung (in 2000), but his 2009 book gave a wildly different orientation. He confirmed that the earlier statement was correct, and thus the 2009 statement was a "mistake" (his word). So, he finally sort of grudgingly acknowledged the mistake. But here, in his 2016 book, we have Aveni returning to the wrong statement, even going further to state that as being *what I had found*.

---

[26] www.Update2012.com/Review-Aveni-Izapa-ballcourt.pdf.

In the very next sentence Aveni's errors continue. He states that I do not "subscribe to conventional interpretations of the Izapa monuments" (204-205). This is false. For Stela 60, Stela 69, Stela 67, MM 25, and Throne 2 and related monuments, I completely subscribe to the basic interpretations of the Brigham Young scholars and other iconographers who have studied the site (Milo Badner, Virginia Fields). I also concur with Timothy Laughton, Dennis Tedlock, Barba Piña de Chan (and others) that the Izapan monuments depict episodes from the Hero Twin Creation Myth, involving the Hero Twins, Seven Macaw, and "First Father" (One Hunahpu). What I have added to the interpretations, as an evidence-based extension of the existing standard readings, is the astronomical orientation data that I have discovered and documented.[27] As such, the ballcourt's winter solstice *sunrise* orientation provides an interpretative basis for deducing that the rebirth of the First Father deity that is portrayed, for example, on Stela 67 — who is acknowledged as a solar deity — represents the December solstice sun. This is logical and is based on *the evidence*. Furthermore, the ballgame itself is about a World Age level of solar rebirth, and the orientation of the ballcourt with its throne indicates which "sun-face" or "day-god" is getting reborn.

Aveni reports wrong descriptions of my work, ignores relevant material, and overlooks the evidence I discovered and was the first to publish (1996, 1998), which naturally factors into my astronomical interpretation of the meaning of the Izapan ballcourt monuments. My augmented interpretations are completely congruent with the accepted dialectic in the Hero Twin Myth (between Seven Macaw and One Hunahpu) and the ballgame symbolism. Seven Macaw is shown on Stela 60 being defeated by the Hero Twins — that interpretation comes from Laughton, Piña de Chan, and Garth Norman (who studied the site with Brigham Young University and did detailed drawings of all the monuments).

Aveni then (still in the same problematic paragraph) emphasizes Julia Guernsey's new perspective that the Izapan characters represent actual rulers at Izapa. *This* was unconventional, as earlier scholarship noted that most of the monuments contain an upper and lower frame representing the open mouth of a snake or jaguar. It is a stylized frame that means 'this scene happens in the Otherworld,' which is to say, that they are essentially *mythological* depictions. It may be that the mythological First Father dialectic with Seven Macaw — clearly a central dynamic in the ballcourt carvings — served as a mythological prototype for the sacrificial obligations of actual rulers at Izapa, and Guernsey's work emphasizes this *unconventional* possibility. That's fine. However, her 2006 book on Izapa, which Aveni cites for the "standard" interpretation of Izapa (which it was not), analyzes only one ballcourt monument! Her book is frequently cited by my critics for the better, more comprehensive, interpretation of Izapa, supposedly obviating my analysis of the Izapa ballcourt monuments, but *she doesn't even examine the*

---

[27] A few of these interpretations might be seen as "unconventional," but that's because they integrate new facts and reflect my consideration of this new evidence, mainly from archaeoastronomy, which previous commentators had overlooked. The earlier "conventional" interpretations are based on incomplete data.

ballcourt.[28] My work remains the most thorough and comprehensive treatment of the Izapan ballcourt, its fifteen stone artifacts and astronomical orientations, and Aveni doesn't acknowledge this. Rather, he crafts a misleading and incomplete send-up.

Aveni's paragraph on my Izapa work contains multiple errors and misleading assertions, after which, to top it off, he claims that my late-stage 2012 strategy was to hedge my bets "like Argüelles" did, just in case "no global transformation took place on December 21, 2012" (205). This shows that his cognitive processing and presentation of my work is distorted and disingenuous. Virtually every statement Aveni makes (pp. 204-205) about Izapa and my Izapa work is factually false or misleading. Furthermore, he simply repeats his previous errors, which were explicitly pointed out to him and his publisher in early 2015, *while he was working on his new book.*

Aveni's misleading assumptions continue. He asserts I am "hostile to critics" (p. 212). No, not all critics. I welcome and have long encouraged informed critique and dialogue. I am hostile to and critical of degreed scholar-critics publishing peer-reviewed articles or books who refuse to acknowledge or correct their factual errors and refuse to adjust their views based on presented evidence, thereby violating science and the principles of sound scholarship. I am hostile to scholars publishing in peer-reviewed journals or books who falsely portray me and my work, repeatedly ignore dialogue and facts, pollute the published record with incorrect falsehoods that are potentially damaging to my work and livelihood, despite my seeking resolution through proper official channels.

Aveni's book contains not only the factual errors I just enumerated, but loose opinions and baseless assertions, much in the way that his 2009 book was (*2012: The End of Time*), which Kevin Whitesides pointed out in his Amazon review.[29] Aveni concludes his Chapter 11 with judgmental fervor, saying that:

> …as long as **the perennialists** choose to turn a deaf ear to the solid evidence that reveals our ancient human ancestors to be as flawed as we, our fantasy-loving, gullible, popular culture will continue to be influenced by **their** artfully crafted "**invented sacred traditions**" (Aveni 2016:185, citing the title of Hammer & Lewis 2008, emphasis added).

This statement recapitulates Aveni's misunderstanding of the Perennial Philosophy as being "invented" (following Hoopes and Hammer) and furthermore ignores 2012 as a valid artifact of ancient Maya thought. I was onto this approach early on and engaged a rational investigation of 2012 back in the early 1990s, at a time when

---

[28] Apart from a picture caption showing Tacana volcano, Guernsey mentions the Izapan ballcourt only once (2006: 172), merely as the location of Stela 67, which she elsewhere (137) interprets as "the transportation of an individual" (a ruler) who is "clasping scepters" while imitating a deity. Echoing *my own earlier interpretation* (1996, 1998), she notes that it "anticipates Classic Maya portrayals of the Maize God's [First Father's] rebirth" (137). She *briefly mentions* Stela 22 (which was found by the road outside the ballcourt).

[29] See also http://update2012.com/ResponsetoAvenisarticle.html. And others at http://Update2012.com.

Aveni and his colleagues considered 2012 to be a joke. He, and several other hold-outs in academia, still depict it as such. But properly understood, my work and that of a few other progressives in Maya Studies, rally evidence that the ancient Maya were aware of the precession of the equinoxes as well as the precessional alignment that culminates on December solstices in the years around 2012.[30]

I am, of course, concerned with my work being accurately treated in peer-reviewed publications by Maya scholars, even while critiques are being offered. My work is, first and foremost, about an evidence-based reconstruction of ancient Maya astronomy and period-ending beliefs. These period-ending beliefs touch upon profound ideas reflected in many religious traditions, and this is where scholarly critics like Aveni bump their heads as they consider my work. They can't seem to get past the fact that the ancient Maya possessed "spiritual teachings," and critics like Aveni dislike that I express admiration and respect for such ideas (such as non-duality, which is stripped down to be "reciprocity" in the clinical terminology of un-philosophical anthropologists).

Meanwhile, that part of my work can be treated separately from my astronomical reconstruction work, which is an interdisciplinary argument integrating evidence from archaeology, astronomy, calendrics, archaeoastronomy, iconography, and Creation Myth symbolism. In the realm of fact-based assessment (which is his turf), Aveni's critiques utterly fail, despite a plethora of clearly written summaries, evidence, and detailed arguments in my work that even High School students can understand and accurately report.[31] Aveni's book is conceptually biased and contains many errors of citation and attribution. He draws from Hoopes's hostile and flawed studies, *which have already been corrected in the peer-reviewed literature.*[32]   Like Hammer, he ignores the primary voices of the Perennial Philosophy. He appropriates and distorts my own definitions and cherry picks quotes and bits of information taken out of context in order to bolster his baseless convictions. The rational and accurate processing of information, even by a degreed officer of the Academy, is quite broken here. Over 135 college and university libraries have rushed to order his book, not knowing or perhaps not even caring about all the conceptual deceptions and factual errors that it contains.

It is curious that the errors in Aveni's book are found almost exclusively in the final four chapters, where my work and the ideas I pioneered are treated. This suggests a prejudicial bias that is rooted in baseless preconceptions or a perhaps unconscious reflex to mitigate or denigrate. This can be explained as the gatekeepers in a field of study trying to exclude an outsider who made valid contributions on a topic (2012 astronomy) that was mishandled and misunderstood by professional scholars until new data (e.g., Tortuguero Monument 6 and La Corona Block 5) forced them to reverse their previous assumption that 2012 was unworthy of serious rational consideration, essentially a non-topic or worse, a joke.

---

[30] Grofe 2011a, 2011b; MacLeod and Van Stone 2012, Jenkins 2010, 2011b, 2014b.

[31] I refer to an impressive paper I received from a High School student named Jack Mazza in 2010, which I have posted on my website: http://alignment2012.com/JackMazza-paper-on-2012.pdf.

[32] Jenkins 2014a.

For many years it could be justifiably said that *Maya scholars never dropped the ball on seriously treating 2012, because they never picked the ball up.*

Aveni's books will reinforce a negative picture of my work and contributions so long as Aveni, his publisher (the University Press of Colorado), the press Director (Darrin Pratt), and Peter Berkery at the AAUP refuse to acknowledge and correct the errors. However, unlike my efforts to correct his 2009 book, this time around all of the errors I indicated were acknowledged by Aveni and his publisher. The exact same procedure and process that didn't work last year, worked this year. This can only be explained by persistence on my part as well as a repeated appeal that they rationally engage the non-negotiable factual nature of the problems, which one assume scholars and their university press publishers should be capable of doing.

But, notwithstanding this recent victory, for many years scholars have constructed hostile judgments of the 2012 topic, while ignoring or misrepresenting evidence-based research and findings. I selected Aveni's books to illustrate something that has consistently infected the attitude toward 2012 within Mayan Studies. He is a pioneer of archaeoastronomy (which is the primary field of evidence for my 2012 work); he is a senior scholar well respected among his colleagues and his conclusions and judgments are often followed without question (e.g., Restall & Solari, 2010; Van Stone 2010; Krupp 2015). The recent corrections to Aveni (2016) require that certain errors in Aveni (2009) be likewise corrected, as well as all the undiscerning citations to Aveni's errors made by his colleagues, who mistakenly thought they were viable critiques of 2012 and my work.

In the comparison between his denial of the errors in his 2009 book and the acknowledgement of all the errors in his 2016 book, we have a rare and valuable lesson to be learned. We have, here, evidence for what was basically a communication problem, where the problem was completely on the end of the receiver — a scholar who could not or would not receive evidence. This is now clearly identified as the problem that has afflicted the accurate reception of my work in Maya Studies. I've made defensible evidence-based arguments, have sent my work to scholars for comment for some twenty years, inviting dialogue, have played the game by publishing in peer-reviewed publications and presenting at academic venues, but have largely received murky dismissals, knee-jerk judgments not based on what I've actually written, and snarky, baseless, character assassination attempts. An uncomfortable situation of cognitive dissonance was thus the unavoidable outcome among stubborn scholarly critics when their colleagues (such as John B. Carlson) began to echo the very same interpretations about 2012 *that I pioneered*, many years ago. It has been a revealing episode in the field of Maya Studies that demanded a clear and informed correction of the published record, which has now, finally, in this particular case, been achieved.

◈

Primary Sources and References Not Detailed in the Foot Notes

Aveni, Anthony. 2009. *2012: The End of Days*. University Press of Colorado.

Aveni, Anthony and Horst Hartung. 2000. "Water, Mountain, Sky: The Evolution of Site Orientations in Southeastern Mesoamerica." In *Chalchihuitl in Quetzalli: Precious Greenstone Precious Quetzal Feather: Mesoamerican Studies in Honor of Doris Heyden*, edited by Eloise Quiñones Keber, 55-65. Lancaster: Labyrinthos.

Callaway, Carl. "Primordial Time and Future Time: Maya Era Day Mythology in the Context of the Tortuguero 2012 Prophecy." In *Archaeoastronomy*, Vol. 24. J Carlson, ed. Stated publication date 2011, released August 2012. University of Texas Press.

Carlson, John B. 2011. "Anticipating the Maya Apocalypse: What Might the Ancient Day-Keepers Have Envisioned for December 21, 2012?" In *Archaeoastronomy*, Vol. 24. J. Carlson, ed. Stated publication date 2011, released August 2012. University of Texas Press.

Coe, Michael. 1966/2015. (9th edition of 2015 w/ Stephen Houston). *The Maya*.

Coe, Michael. 1988. *Mexico*. 3rd edition, revised and enlarged. Thames & Hudson.

Freidel, David. 2009. Comments in a CNN Interview and my response. See: www.update2012.com/May2009.html and www.update2012.com/response-to-freidelMay.html.

Freidel, David A. and Marcos Adrián Villaseñor. 2009. "Ancient Maya Prophecy and Calendrics: Reality and Fantasy." Online piece with my rebuttal "Requiring That Ancient Maya Stargazers Be Like Modern Astronomers:
The Polemical Sophistry of Marcos Villaseñor and David Freidel", both online here: www.update2012.com/ResponsetoVillasenor.html

Grofe, Michael. 2011a. "Measuring Deep Time: The Sidereal Year and the Tropical Year in Maya Inscriptions." In *Ethnoastronomy and Archaeoastronomy: Proceedings from the Ninth Oxford International Symposium on Archaeoastronomy*. C. Ruggles, ed., pp. 214–230. Cambridge: Cambridge University Press.

—. 2011b. "The Sidereal Year and the Celestial Caiman." In *Archaeoastronomy*, Vol. 24. J. Carlson, ed. Stated publication date 2011, released August 2012. University of Texas Press.

—. 2012-2013. "The Copan Baseline and the Outlier Stela." *Archaeoastronomy*, Vol. 25. J. Carlson, ed. Stated publication date 2012-2013, released March 2015. University of Texas Press.

Guernsey, Julia. 2006. *Ritual and Power in Stone*. University of Texas Press.

Hammer, Olav. 2001. *Claiming Knowledge: Strategies of Epistemology from Theosophy to the New Age*. Brill Publications.

Hammer, Olav and James R. Lewis (eds). 2008. *The Invention of Sacred Tradition.* Cambridge University Press.

Hoopes, John. 2011. Review of the 2012 books of Aveni and Van Stone. *Archaeoastronomy Journal*, Vol. 22. University of Texas Press.

Hoopes, John and Kevin Whitesides. 2014. Response to John Major Jenkins. In *Zeitschrift für Anomalistik.* Band 14, Nr. 1. The Society for Anomalistiks. Germany.

Isbell, Billie Jean. 1982. "Culture Confronts Nature in the Dialectical World of the Tropics." In *Ethnoastronomy and Archaeoastronomy in the American Tropics,* edited by Anthony Aveni and Gary Urton, pp. 353-364. Annals of the New York Academy of Sciences, Vol. 385. New York.

Jenkins, John Major. 1989. *Journey to the Mayan Underworld.* Four Ahau Press. 2nd edition 2017: www.Amazon.com.

—. 1992. *Tzolkin.* Four Ahau Press. Reprinted with BSRF 1994.

—. 1996. *Izapa Cosmos.* Four Ahau Press.

—. 1998. *Maya Cosmogenesis 2012.* Bear & Company.

—. 2002, *Galactic Alignment.* Inner Traditions International.

—. 2007. "The Origin of the 2012 Revelation." In *The Mystery of 2012.* Louisville, CO: Sounds True.

—. 2009. *The 2012 Story.* Tarcher / Penguin Books.

—. 2010. "Astronomy in the Tortuguero Inscriptions." Paper presented at the 75th Meeting of the Society for American Archaeology. Paper posted at *The Maya Exploration Center* and http://thecenterfor2012studies.com/Astronomy-in-TRT-SAA.pdf.

—. 2011a. "The *Maya Exploration Center* Facebook Discussion on the Astronomy of 2012 and Tortuguero Monument 6." Published at the *Maya Exploration Center* and http://thecenterfor2012studies.com/MEC-Facebook-Discussion-2010-ON-Jenkins-SAA-TRT-Astronomy.pdf.

—. 2011b. "Approaching 2012: Modern Misconceptions versus Reconstruction Ancient Maya Perspectives." In *2012: Decoding the Countercultural Apocalypse.* J. Gelfer, ed. Equinox Publishing, Ltd.

—. 2012. *Reconstructing Ancient Maya Astronomy.* Four Ahau Press.

—. 2014a. "The Coining of the Realm (of the 2012 Phenomenon): A Critique of the Whitesides and Hoopes Essay." In *Zeitschrift für Anomalistik.* Band 14, Nr. 1. The Society for Anomalistiks. Germany.

—. 2014b. "Lord Jaguar's 2012 Sacrifice: Astrotheology and Magical Invocation in a 7th-century Maya Inscription." *Clavis Journal.* Three Hands Press.

—. 2015. "Review of the 9th Edition of Michael Coe's *The Maya* (released June 2015)." At: http://thecenterfor2012studies.com/Coe9-2015.pdf.

—. *Ivory Tower, House of Cards*. n.d. (written 2015-2016).

Krupp, Ed. 2015. Preface to *Cosmology, Calendrics, and Horizon-Based Astronomy in Ancient Mesoamerica*. S. Milbrath and A. Dowd, eds. University Press of Colorado.

MacLeod, Barbara. 2008. "The 3-11-Pik Formula." Paper presented at the Maya Hieroglyphic Meetings in Austin, Texas, March 2008. Posted with the author's permission: http://alignment2012.com/3-11PikFormula.html.

—. 2011. "Holding the Balance: The Role of a Warrior King in the Reciprocity between War and Lineage Abundance on Tortuguero Monument 6." *Archaeoastronomy*, Vol. 24. J. Carlson, ed. Stated publication date 2011, released August 2012. University of Texas Press.

MacLeod, Barbara and Mark Van Stone. 2012. "The Great Return." *Zeitschrift für Anomalistik*. Band 12. The Society for Anomalistiks. Germany.

Meeus, Jean. 1997. *Mathematical Astronomy Morsels*. Willman-Bell.

Restall, Matthew and Amara Solari. 2011. *2012 and the End of the World*. Rowman & Littlefield Publishers.

Rice, Prudence, 2007. *Maya Calendar Origins*. University of Texas Press.

Sitler, Robert. 2006. "The 2012 Phenomenon." *Nova Religio*. Vol. 9, No. 3. The University of California Press.

—. 2010. *The Living Maya*. North Atlantic Books.

Stray, Geoff. 2005. *Beyond 2012: Catastrophe or Ecstasy – a Complete Guide to End-Of-Time Predictions*. East Sussex, U.K.: Vital sings Publishing.

Stuart, David. 2011. *The Order of Days: Unlocking the Secrets of the Ancient Maya*. Three Rivers Press.

Tedlock, Barbara. 2005. Comments (in a transcript of the interview for the film): http://www.nightfirefilms.org/breakingthemayacode/interviews/TedlocksTRANSCRIPT.pdf. From *Breaking the Maya Code*, Night Fire Films.

Van Stone, Mark. 2010. *2012: Science and Prophecy of the Ancient Maya*. Tlacaelel Press.

Wallace, Patrick. In Jenkins (2002)

Whitesides, Kevin. 2015. "2012 Millennialism Becomes Conspiracist Teleology." *Nova Religio*. Vol. 19, No. 2. The University of California Press.

Whitesides, Kevin and John Hoopes. 2012. "Seventies Dreams and 21st-Century Realities. In *Zeitschrift für Anomalistik*, Band 12. The Society for Anomalistiks. Germany.

—. 2014. "Mythology and Misrepresentation: A Response to Jenkins." In *Zeitschrift für Anomalistik*. Band 14, Nr. 1. The Society for Anomalistiks. Germany.

Addendum Update: Did it Happen Yet?

September 29, 2016. After waiting a month for any news from the publisher that the promised corrections were being added to the second printing and the eBook, I made a simple inquiry. I received a response that the press Director (Darrin Pratt) didn't know, and would ask his production department. The next day I received a compiled list of the errors that were incorporated into the revised book, and a PDF of the revised book. The "last save" date-stamp on the press-ready PDF was the previous day, so I suspected they got on this as a consequence of my inquiry. In briefly comparing the two documents I immediately noted that a few things were omitted, which I'd pointed out in emails to Darrin while the process was unfolding in July and August. To re-engage and re-check the material in Aveni's book seemed an onerous re-immersion into the toxic soup, and I was just finishing some editing work on a friend's book, eyes bleary, so I set the latest challenge aside for a few days. I then engaged it with focus and found that perhaps the most important correction had been overlooked. I composed the following corrections and sent them to Darrin:

Cover email:
Hi Darrin,   September 21, 2016
Thank you for including me in the process of corrections. I think you'll be happy that I caught a few things, including one item where one of Aveni's requested corrections was actually not done in the corrected PDF you sent me, which resulted in an incorrect end note sequencing. I have very carefully explicated the needed corrections in clear language, but feel free to contact me if you have any questions. I trust these additional items will be helpful and will be added.

Also, I think it would be important to distinguish this corrected second printing by added a phrase to the front matter, something like "Corrected second print, September 2016." Can this be done? Best wishes,

John Major Jenkins

Corrections File (pasted into email and attached as an MSWord document:

Dear Darrin,   September 21, 2016

I am going on a combination of the correction list you sent me on 9-14 and what is represented in the revised PDF of the book that you sent me, as well as Aveni's original emails with his corrections. There are several points that have not been entered into the PDF correctly, according to what Aveni himself stated, and none of my corrective points have been included. Therefore, I went back through our email exchanges to find the pertinent points.

I refer to pagination in the PDF you sent. Note 11 on page 234 includes the page reference "94" which is indeed what Aveni stated should be added to correct the originally erroneous citation. However, as I indicated in an email to you (August 1) this page in Argüelles's book is a full-page diagram with no text and is meaningless as a page reference for the passage that Aveni quotes. The correction was achieved by also citing the correct page for the quote, 184, and that is all that is needed. So it's just a matter of removing the p. 94 reference. So, in note 11 on page 234, remove page 94 from the reference:

Argüelles, *The Mayan Factor* (Santa Fe: Bear & Co. 1987), ~~94,~~ 184.

The second correction in your email of 9-14 was entered, in the PDF, on the top of page 196:

p. 196, l. 1: —we will reconnect the the "psychically received" heliotropic octaves" in the "solar activated magnetic field."

Two problems here. In the corrected PDF, your editor replicated the double "the" so you might want to remove one of them. More importantly, Aveni added "psychically received" to this quotation string, when in fact it needs to be added to the quotation at the top of page **197**. I suggested in an email how Aveni's passage (on page 196) could be re-structured but the point is that "psychically received" does not belong there. Instead, it needs to be inserted into the Argüelles quotation at the top of page **197**, which apart from this has been correctly revised. From the PDF:

As José Argüelles explained: "Amidst festive preparation and awesome galactic-solar signs ***psychically received***, the human race, in harmony with the animal and other kingdoms and taking its rightful place in the great electromagnetic sea, will unify as a single circuit. Solar and galactic sound transmissions [Is this perhaps a relic of the McKenna brothers' experiments?] will inundate the planetary field. At last, Earth will be ready for the emergence into interplanetary civilization."12

I added the phrase (in ***underlined bold italic***) where it needs to be inserted. That is the correct quotation from Argüelles, which end note 12 now correctly cites to Argüelles 1987: 194.

Now, returning to the incorrect placement of that phrase <u>on the top of page 196,</u> and the extra "the." If those are simply removed then the corrected passage will read:

All our senses will attain new revelations, for then, Argüelles tells us in typical science jargon, we will reconnect the "heliotropic octaves" in the "solar activated magnetic field."11

But now it is slightly nonsensical as to Argüelles's full passage, but you can add "with" to accurately reflect his meaning:

All our senses will attain new revelations, for then, Argüelles tells us in typical science jargon, we will reconnect **_with_** the "heliotropic octaves" in the "solar activated magnetic field."11

Now, A Very Important Correction:
I noticed that one of Aveni's final corrections, which he sent to both you and I on August 13, was <u>not added</u> to the "corrected" PDF. From Aveni's correction-list of August 13:

p. 202, l. 27:

Delete: "As anthropologist — pointed out, many of the ideas about post-millennial prophecy draw on the work of — (old note 35)"

Delete old note 35, p. 235.

This is also stated in the list of corrections you recently sent me, on 9-14. The *entire passage* from beginning to end, *and its end note*, were to be deleted. The sequence of end notes within the text were to be preserved, because Aveni also requested that a <u>new note 35</u> be added, just prior to this now deleted passage & its end note. Also, with the old note 35 (the John Hoopes source) deleted from the End Notes section, and the new end note 35 added there (a citation to my book *Galactic Alignment*), the sequence of the notes in the End Notes would have been likewise preserved. However (a BIG however here), while your editor DID add the new note 35, your editor neglected to delete old note 35, renamed it note 36, and adjusted the remaining sequence of end note numbers (in the End Notes section on pp. 235-236). The editor also adjusted, by 1, the sequence of superscripted end notes in the text, following end note 35. These resequencings were unnecessary, as Aveni had cleverly accounted for this.

So, what needs to happen here is that, on page 235 of the "corrected" PDF you sent me, **_note 36 "Hoopes, Mayanism Comes of (New) Age"_** needs to be deleted. Then the following sequence of end notes needs to be restored to its original sequence, such that the final end note is **_#51 Godwin, Atlantis and the Cycles of Time_** on page 236. In addition, the sequencing within the text needs to be restored, as your editor increased all the superscripted end note values by 1, beginning with the Schele & Freidel citation (line 19 of page 203). In the PDF this needs to be changed from 37 to 36, and all subsequent superscripted end notes in the text need to be restored to their original sequence, ending in 51 (p. 206, line 5).

If Aveni's clever addition of new note 35 and his deletion of old note 35 (on page 202) would have been correctly adopted, these complicated end note numbering changes and re-corrections would have been unnecessary. This one is rather critical because Aveni had correctly requested the deletion of the erroneous Hoopes source, but that didn't get performed in the "corrected" PDF that you sent me.

Finally, there's some messy formatting with the new passage that Aveni requested be added after end note 34 on page 202 (lines 26-27). You need a space after the superscripted end note 34, and the second "i" in the book title *Galactic Alignment* needs to be italicized. Also, there is no need for the superscripted end note 35 to be italicized.

In the mid-1980s I wrote and edited step-by-step instruction manuals for kits. Since the late 1980s I have edited hundreds of books for twelve different publishers and I worked at ILE and netLibrary in Boulder the 1990s-2001, not to mention my own books and essays, so I know how to spot and explicate these kinds of mix-ups, which often happen in publishing. I hope you understand that my comments are geared toward helping you correctly apply *what Aveni himself requested and approved.* In two instances (the first two I described above) there's a typo and a mistaken phrase placement that Aveni certainly did not intend, and which I've noted in the interest of the book being more accurate. I think I've been surgically clear in my comments, but please contact me if you have any questions. Best wishes,

John Major Jenkins

Darrin responded the next afternoon, thanking me for double-checking and he explained that his production department may have gotten Aveni's agreed-to revisions out of context while compare the different lists and comments. Nevertheless, he assured me they would get fixed, and said there was a process for indicating a second printing (though he didn't say a *revised* or significantly *corrected* second printing). He said he would look into it. I responded with a brief "Okay, great. Thank you and best wishes, John."

So, apparently we are one step further, six weeks later, to the corrections actually being enacted. The process began with my initial query on June 15, three-and-a-half months ago. I just re-checked the Worldcat.org database to see how many more of Aveni's books were now on the shelves of libraries (mostly academic) around the world. The number is now up to 149, which is about 25 more than when I first checked in July. A few of the listings include the sub-title "From Millerism to Mayanism" and this indicates what was probably an early working sub-title supplied by the publisher in pre-press announcements to purchasers at libraries.

Final Update. October 5, 2016. Darrin sent me the final revised and corrected PDF yesterday. I looked through it and confirmed that my latest notes and corrections (detailed above) were incorporated into it. In addition — and this is an important point — a "Corrected second printing" note was added to the front matter. This PDF is the colorized eBook file, which I assume will also be used as the printer's file, for the second printing, when that happens. It should happen soon, as several months ago I was informed that only about 200 copies of the first printing remained in stock. As of today, there are 149 copies listed on Worldcat.org, all from the first printing, and ten of them are at college and university libraries in my state, Colorado (including the universities in Fort Collins, Greeley, and Boulder), and one in nearby Wyoming. Some of these first edition copies are at libraries in Europe and around the world. So, the mitigation mission was largely accomplished.

Since I am a member of the *Society for American Archaeology*, the SAA, (or at least was in the 2010-2011 presentation season), and gave a presentation at SAA 2010, I inquired at the *Latin American Antiquity* journal (which is published by the SAA) regarding my submission of a lengthier review-essay of Aveni's book. This possibility was suggested to me by Dr. Robert Benfer. There are two other options for publication I'm pursuing. A minimalist option is to post this current lengthy review on Academia.edu (where I have some of my other essays posted) and then submit a brief "notice" to the selected journals, listing the corrected errors and with a brief accounting of the process. There is already at least one review of Aveni's book out there (August 2016), which like many academic reviews is superficial and undiscerning. It missed all the errors, and we can expect that virtually every other "in house" review of Aveni's book will do likewise. That is precisely why a longer review-essay is necessary, written by someone who is knowledgeable about all the facts and issues. And this one is unique because it documents the painstaking process by which many factual errors were successfully identified, corrected, and approved by the author, and the second printing was revised according to the principles of academic publishing.

October 7, 2016. The *Latin American Antiquities* journal requires that you sign up to be a member before submitting manuscripts and reviews. I did this, but before submitting my review of Aveni's book, I inquired:

Greetings,   October 3, 2016
As a member of SAA I gave a presentation at the 2010 SAA meetings in the Archaeoastronomy forum moderated by Dr Robert Benfer. He recently recommended to me that I submit my book review to you, but it is more along the lines of a longer review-essay which includes corrections to the book that are going into a revised second printing of the book. Consequently, it is an important record of the already approved corrections with a discussion of the implications which are not explicitly dealt with in the revised book. The book was first published in May of this year (2016) with the University Press of Colorado — I offer a review plus a discussion of the author-approved errata. I would like to speak with someone regarding the allowable parameters (length) for this special circumstance before I finalize my submission. I have ready created a user account for your online submissions program. Thank you and best wishes,
    John M Jenkins

I received two responses from two different people, one of which stated that they didn't accept book reviews that weren't solicited. The other person forwarded my query to someone else for comment but, so far, no response (and it never happened). One other academic think-tank and publication venue online has declined to post my review-essay. Other academic journals (such as *Nova Religio*) have strict word-count limitations and guidelines which can only result in laudatory puff-pieces rather than meaningful critical reviews of academic publications.

Online Appendices

Appendix 1. The Rest of the Iceberg
        (separate file, summarized here)
  a. Stuart-Houston
  b. Campion
  c. Van Stone
  d. Stan Guenter (see: www.update2012.com/Demonstration-for-Guenter.pdf)
  e. Review of MacLeod & Van Stone's "Great Return" article, with my response to their invitation/challenge
  f. Review of Michael Grofe's 2003 article
  g. Various Others (12 topics)

Appendix 2: Additional Online Resources
  a. Annotated List of Links
  b. The Center for 2012 Studies
  c. Update2012.com

Appendix. 3: Verbatim Filed Complaints and Emails
        (Separate book, summarized here)

Description of Appendices

In order to save paper, Appendix 1 and Appendix 3 will be summarized briefly below and linked online (see p. 156 and p. 158).

In Appendix 1, I address a few ancillary components that I decided not to include in the main narrative of the book. As it is, the material and the events covered here must be overwhelming enough for the reader. But there are three that I'd like to include here, to augment the book. Stephen Houston and David Stuart led the charge to proclaim that 2012 meant absolutely nothing to the Maya. I share comments on their blog, various exchanges, and my review of Stuart's "2012" book, published with a trade publisher in June of 2011. Houston's role in revising the 9th edition of Coe's book *The Maya* is covered in section 7 of Appendix 3 (see below).
    My communication with Nicholas Campion illustrates how a scholar looking into Mayas Studies, from the outside as it were, internalizes the climate of mitigation and seeks legitimization by echoing the prevailing attitudes. Arriving at this position does not necessarily involve careful discernment. My question to Campion was a simple one. Given that several other scholars (Carlson, Callaway, MacLeod) assumed and accepted that there was some kind of "Maya prophecy" in 2012,

worthy of exploration and articulation, why was my work connected with his MPM container (the "Maya Prophecy Movement"), whereas the other scholars were not? My work has been concerned with reconstructing ancient Maya ideas and beliefs, just like the others. And, ironically, my pioneering ideas of the 1990s came to be echoed, much later, by those very same "real" scholars who weren't rounded up for inclusion in the MPM. Campion, after three months and several persistent email exchanges, never answered my simple question.

Appendix 1 also includes a few examples from the work of Mark Van Stone — his self-published book of 2010 and two of his articles (one of which was a prize-winner, co-written with Barb McLeod and published in mid-2012). It also includes a review of the 2011 book by Matthew Restall and Amara Solari, and a variety of other items. Appendix 1 will be posted for free online at: http://www.update2012.com/app1-IvoryTower.pdf.

In Appendix 2, which is here in full, my intention is to provide links to additional essays, reviews, interviews, and so on. I recently updated *The Center for 2012 Studies* website, which I founded in 2010. On the front page I've added "featured" links at the top: http://thecenterfor2012studies.com. In addition, in late 2014 I updated my website called Update2012 (http://www.update2012.com), which I founded in early 2009. It is now an exhaustively comprehensive resources which reviews scholarly writings on 2012, attacks on my work, and related behaviors. If *Ivory Tower, House of Cards* merely whets your appetite for more (much more!) of the same, then this free online resource is for you. My website domain for my book *The 2012 Story* had to be cancelled (you can't keep them all going forever). But the whole site is reproduced at www.alignment2012.com/the2012story/index.html. And good ol' Alignment2012! This was my primary website, first launched in September 1995. There are actually three previous versions of it, accessed through the "News and Updates" tab. This site more or less went static after 2012. It is still a valuable resource for many free articles, reports, and reviews: http://www.Alignment2012.com. Finally, my personal website has become sort of an umbrella for everything else, with a focus on my poetry, new book releases, and letterpress printing: http://JohnMajorJenkins.com.

Appendix 3 provides the indispensable documentation for this book. It contains the verbatim complaints and exchanges with scholars, and it is detailed and exhaustive. Here is the Table of Contents:

1. Anthony Aveni, University Press of Colorado, the AAUP
   a. Timetable of communications
   b. First round of emails with University of Colorado Press director, Darrin Pratt
   c. Complaint with Aveni's errors sent to Pratt
   d. Second round of emails with Pratt (May 2015)
   e. Email queries sent to AAUP Committee for Member Standards and Policy
   f. Cover letter with Complaint files forwarded
   g. Ensuing emails and phone calls, through change of committee chair and members

h. Their "decision," and ensuing communications with the Executive Director of the AAUP, Peter Berkery

2. David Morrison, NASA, and the NASA Communications Policy Office
   a. First Complaint Request sent, re Morrison's comments on the NASA website, sent to David Weaver, director of the NASA Communications Policy Office
   b. Ensuing emails and phone calls with the Office
   c. Second phase: email contact with Morrison, and I send him my Second Complaint Request, re his comments in his presentations
   d. Ensuing emails and phone calls to Morrison and the others, with no response
   e. Timetable of communications

3. Ed Krupp, Griffith Observatory, *Sky & Telescope*, The Beckman Center
   a. Final letter sent to Krupp during correspondence in the 1990s
   b. Emails with Krupp, 2015
   c. Cover letter for package of writings sent by mail
   d. Corrections and questions sent, with invitation to respond
   e. Confirmation of receipt, then silence

4. John Hoopes, University of Texas Press, Wikipedia, *Zeitschrift für Anomalistik*
   4a. How it Started: Hoopes's Refusal to Answer a Simple Question
   4b. Hoopes Makes the Maya World Age Doctrine a New Age 2012 Mythology
   4c. Calling out Hoopes, July 2011
   4d. Hoopes Evades Discussion on Whitesides' FB Page
   4e. I Respond to Hoopes's Questions, March 2011
   4f. Hoopes's Mayanism (from *The 2012 Story*)
   4g. Hoopes's Mayanism Prison (www.alignment2012.com/Mayanism-John-Hoopes.pdf)
   4h. 2011: Complaint filed with University of Texas Press, re AJ Vol. 22
   4i. Ensuing emails with *AJ* editor John B. Carlson (Hoopes cc'd)
   4j. My *Zeitschrift für Anomalistik* review of the Whitesides-Hoopes article
   4k. Ensuing email exchanges with Hoopes (and Whitesides elsewhere)
   4l. My Response to their Response (August 2014)

5. John B. Carlson, Robbins Museum, *Archaeoastronomy Journal*, University of Texas Press, University Press of Colorado
   a. Article proposals & queries to Carlson and *AJ*, 1994-1997
   b. Emails with Carlson, re his Robbins Museum talk in 2010, and related emails
   c. Transcribed Excerpts from Carlson's 2010 Talk
   d. Cordial Email with info links and query, sent to Carlson in 2012
   e. Email sent to Carlson in June 2015, re Milbrath and his UP of Colorado article

6. Susan Milbrath, University Press of Colorado (Clemency Coggins, John B Carlson, Anthony Aveni, Ed Krupp)
   a. Exchange with Milbrath in IMS newsletter, 2007-2008
   b. Email exchange with Coggins
   c. Cover letter with review of the anthology, sent to editors Milbrath and Dowd

d. Ensuing emails with Milbrath

e. Report sent to University Press of Colorado director Darrin Pratt

7. Other Files:

a. My review of Michael Coe's 9th Edition of *The Maya*

b. My 1995 review-essay of Houston & Stuart (1994)

c. Letter to Robert Hall (1994), sent to Carlson in 1994

It's important to have these things documented, for truth and honesty and for correcting the published record on everything that has not been accurately presented. The online appendices are at: www.update2012.com/app1-IvoryTower.pdf and www.update2012.com/app3-IvoryTower.pdf. Also see the main page at www.thecenterfor2012studies.com.

John Major Jenkins
Windsor Colorado
November 7, 2015

○ ⊙ ⇧ ⇧ ⇧ ⊙ ○

Note, regarding Anthony Aveni's published correction to his misunderstanding of Grofe's work (see Chapter 3b for discussion): In the "revised" eBook edition of his 2009 book *2012: The End of Time*, released May 2015, Aveni added a note on page 173 in which he dubiously claims he noted his mistake before Grofe contacted him about it, doesn't offer a correct interpretation, and, shifting emphasis off the implications of his error, he concludes that Grofe's interpretation of the Dresden Codex is wanting, based on the Brickers' epigraphic dismissals (in their 2013 book on Maya astronomy). Furthermore, Aveni neglected to acknowledge and cite Grofe's subsequent articles in which he challenged the critiques of both Aveni and the Brickers. All in all, a pathetic half-hearted attempt to do the right thing, which in fact introduced an additional misreading.

The Brickers took issue with Grofe's reading of the Serpent Series initial Distance Number interval because their own interpretation assumes that the periods without coefficients were to be read as "zero", thereby leading to an unstated and very remote date that conformed to their belief that the entire Serpent Series deals only with 91-day adjustments (these being concerned only with tracking the Tropical Year). Both Grofe and an epigraphic colleague who assessed the Brickers' work conclude that their epigraphic readings and assumptions are not supported, and they base their interpretations on too many assumptions. By favoring their own dubious reading, the Brickers dismiss the interpretation offered by Grofe which, in fact, more cogently integrates the epigraphic clues and is thus better supported by the evidence within the text.

Where Aveni stands on this is not from the vantage of him actually examining Grofe's arguments, but merely parroting the opinion of his friends, the Brickers. His insistence on continuing to cast misleading aspersions was maintained even while grudgingly offering a long-overdue "correction" to his misreadings of Grofe's work and the directional motion of the precession of the equinoxes.

Note. On June 8, 2016 I discovered that in May (2016) Aveni had published yet another book with the University Press of Colorado. Titled *Apocalyptic Anxiety*, it uses the 2012 episode as the culminating example of "America's obsession with apocalypse," with four chapters devoted to skewering my work and related ideas. I produced a 29,000-word review-essay, an 8800-word version, and a 2500-word version. Two of these were added to Appendix 1 and the other is included above (as an addendum) in the present book as a late addition to this study. This victory indicates that a sea-change may finally be happening in Maya Studies — a change from ignoring facts to recognizing and accepting facts.

About the Author

John Major Jenkins
(summer 2014)

John Major Jenkins is a pioneering voice in the evolving discussion of Maya cosmology and 2012 with over twenty-five years of experience defining and debating the issues. Informed by innovative field work at key archaeological sites and inspired by living and working among the Highland Maya, Jenkins' comprehensive work covers media misconceptions, assessments of 2012 theories, consciousness studies, Maya shamanism, archaeoastronomical research, Perennial Philosophy, academic misconceptions, and the crisis of sustainability faced in the modern world. His own unprecedented "2012 alignment theory" is now receiving new support from recent discoveries in the Maya inscriptions. While integrating the scientific and spiritual viewpoints, Jenkins articulates and honors contemporary Maya calendar tradition and the ancient Maya vision of a unified cosmos.

Since the conclusion of 2012, John has continued to address and correct academic shenanigans and professional violations in treating 2012 as a valid topic of rational inquiry. Some 800 pages of updates on his websites document the process that has unfolded in the four years since December 21, 2012. The results are in. Science, as it applies to understanding 2012, is broken. Why? Because scientists broke it. NASA and University Press publishers are complicit in broadcasting and publishing false and defamatory statements, and then defending them when confronted with their ethical violations. This situation is especially tragic, because ancient Maya spiritual wisdom and teachings address the intractable crisis that the modern world finds itself in. John's work explores the ongoing, and largely unrecognized, role of the galactic alignment in the vicissitudes of human history, including new archaeological, astronomical, and geological discoveries dating to 12,800 years ago. Jenkins' background includes:

- Advisory director and founding member of *The Maya Conservancy,* a non-profit foundation dedicated to education and the preservation of ancient Maya sites
- National Fellow member of *The Explorer's Club*
- Director of *The Center for 2012 Studies*
- Manager of Update2012.com and Alignment2012.com
- Member of *The Society for American Archaeology* and *The Institute of Maya Studies*

John's work has been featured since 1998 in media produced by ABC *Nightline,* the *U.S. News and World Report,* the *New York Times,* National Geographic, Discovery Channel, The History Channel, and NBC's SyFy Channel. He has taught at numerous institutes and universities nationally and abroad, including the Universidad Francisco Marroquin in Antigua

Guatemala, the Esalen Institute, Kingsley Hall in London, the Society of Henry XIII in Belgium, the University of Southern Oregon, the New England Antiquities Research Association, the Institute of Maya Studies in Miami, the Society for American Archaeology, and Naropa University. He is also a regular at popular venues such as the Whole Life Expo, Megalithomania, Earth Keepers, and the Mind Body Spirit Expo.

Since the 1980s John's articles have appeared in magazines, newspapers, websites, peer-reviewed journals, and book anthologies, including: *Zeitschrift für Anomalistik* (2014), The Institute of Maya Studies *Explorer* (many items 1995-2017), *The Mystery of 2012* (2007, Sounds True), *You Are Still Being Lied To* (2009, Disinformation Company), *Towards 2012* (2008, Penguin), *New Dawn* Magazine (Australia), *Maya Exploration Center* (Research papers, 2011), *Society for American Archaeology*, 75th Conference (2010), *Clavis* Journal (2014), and many others. His major books, translated into twelve languages, include *Journey to the Mayan Underworld* (1989 / 2017), *Tzolkin: Visionary Perspectives and Calendar Studies* (1992/1994), *Maya Cosmogenesis 2012* (1998), *Galactic Alignment* (2002), *Pyramid of Fire* (with Marty Matz, 2004), *Unlocking the Secrets of 2012*, (3-CD audio program 2007), *The 2012 Story* (2009), *Lord Jaguar's 2012 Inscriptions* (2011), *Reconstructing Ancient Maya Cosmology* (2012, based on a presentation given at the New England Antiquities Research Association), and *Three Plumes of Judas* (a novel, 2017).

Websites:
http://JohnMajorJenkins.com
The Center for 2012 Studies: http://TheCenterfor2012Studies.com
Maya Cosmology and Calendrics: http://Alignment2012.com (launched in 1995)
Update2012: http://Update2012.com
The2012Story.com: now at http://alignment2012.com/the2012story/index.html

☼